# Structural Engineering

# Structural Engineering

Wyatt Kelly

Larsen & Keller
www.larsen-keller.com

Structural Engineering
Wyatt Kelly
ISBN: 978-1-64172-113-4 (Hardback)

# ☰ Larsen & Keller

Published by Larsen and Keller Education,
5 Penn Plaza,
19th Floor,
New York, NY 10001, USA

**Cataloging-in-Publication Data**

Structural engineering / Wyatt Kelly.
    p. cm.
Includes bibliographical references and index.
ISBN 978-1-64172-113-4
1. Structural engineering. 2. Engineering. 3. Architecture. I. Kelly, Wyatt.
TA633 .S87 2019
624.1--dc23

For more information regarding Larsen and Keller Education and its products, please visit the publisher's website www.larsen-keller.com

# Table of Contents

**Permissions**

**Index**

# Preface

Structural engineering is a branch of civil engineering that is concerned with the calculation and analysis of strength, stability and rigidity of concrete structures. It uses the techniques and principles of applied mechanics, mathematics and materials science to study how structures support their own weight along with imposed loads. An understanding of the structural performance of different materials and geometries is vital for the construction of structural systems. The creative manipulation of resources, materials and structural elements is an important dimension of this field. Specializations in structural engineering may exist for particular structures such as building, earthquake engineering, civil engineering, mechanical structures, pipelines, tunnels and bridges, among others. This textbook is a compilation of chapters that discuss the most vital concepts in the field of structural engineering. Different approaches, evaluations, methodologies on structural engineering have been included in this book. This book will serve as a reference to all structural and civil engineers, architects and students.

A foreword of all chapters of the book is provided below:

**Chapter 1**, Structural engineering is a branch of civil engineering. It is concerned with the structural performance of different geometries and materials, and calculation of the strength, rigidity and stability of built structures. This chapter introduces in brief about the principles of structural and architectural engineering as well as the varied aspects of earthquake-resistant structures; **Chapter 2**, A study of various building materials and their properties is vital to the understanding of how these materials support structures and resist loads. Some of the common structural materials are concrete, iron, timber, alloys, aluminium, composite materials, etc. This chapter has been carefully written to provide an extensive understanding of common structural materials; **Chapter 3**, Every engineering structure is made up of many structural elements, such as columns, arches, beams, plates, shells, etc. These elements are used to build complex structural systems. The aim of this chapter is to explore the varied aspects of structural elements and systems, such as rigid frame, braced frame, space frame, eyebar, tensile structure, etc.; **Chapter 4**, An understanding of how structures resist and support imposed loads and self-weight is vital to structural engineering. Structural analysis, theoretical and empirical design codes and corrosion resistance of structures and materials are other aspects intrinsic to this field. The topics elucidated in this chapter on stiffness, bending stiffness, deflection, strength analysis, etc. aid an extensive understanding of structural engineering; **Chapter 5**, The determination of the influence of loads on structures and their components is known as structural analysis. Computation of a structure's deformations, stresses, stability, internal forces, etc. are also within the scope of this area. This chapter extensively covers the varied tools and methods used in structural analysis, such as direct integration of a beam, Castigliano's method, Macaulay's method, limit state design, moment distribution method, etc.

At the end, I would like to thank all the people associated with this book devoting their precious time and providing their valuable contributions to this book. I would also like to express my gratitude to my fellow colleagues who encouraged me throughout the process.

**Wyatt Kelly**

# Structural Engineering: An Introduction

Structural engineering is a branch of civil engineering. It is concerned with the structural performance of different geometries and materials, and calculation of the strength, rigidity and stability of built structures. This chapter introduces in brief about the principles of structural and architectural engineering as well as the varied aspects of earthquake-resistant structures.

Structures are any system that resists vertical or horizontal loads. Structures include large items such as skyscrapers, bridges, and dams, as well as small items such as bookshelves, chairs, and windows. Most every day "structures" are "designed" by testing, or trial and error; while large, unique, or expensive structures that are not easily tested are generally designed by a qualified structural engineer using mathematical calculations. Most practicing structural engineers design and analyse buildings, bridges, power plants, electrical towers, dams, and other large structures that are essential to life as we know it.

Structural Engineering is a specialty within Civil Engineering. Structural Engineers create drawings and specifications, perform calculations, review the work of other engineers, write reports and evaluations, and observe construction sites.

The basic tasks of structural engineering relate numerical quantities of physical forces to physical configurations of force-resisting elements. Analysis is the process of determining forces in each element in a structure (such as a beam) when the configuration of elements is already defined. Design is the process of configuring elements to resist forces whose values are already known. Analysis and Design are complementary procedures in the overall process of designing new structures. After performing a preliminary design, the designer estimates the final configuration of elements of a structure, but only until an analysis is performed can the forces in those elements be known. After performing an analysis, the element forces are known, and the elements can be designed (their configuration can be chosen) more precisely. The process iterates between analysis and design until convergence is achieved.

## Architectural Engineering

Architectural engineering is the application of engineering principles and technology to building design and construction. Architectural engineers work together with architects and civil engineers, but are unique in both their skills and role as part of the building design team.

Architectural engineering is a relatively new licensed profession, emerging in the 20th century as a result of the rapid technology advancement of the Industrial Revolution. Architectural engineers are the engineers that specialize in making buildings. This is a particularly important profession today, since our world is immersed in two major technological revolutions, (1) that of rapidly advancing computer-technology, and (2) the parallel revolution arising from the need to create a sustainable planet. Architectural engineers are at the forefront of both historical opportunities.

Built on a legacy of thousands of years of gradual innovation in construction technology and scientific advancement, architectural engineers apply the latest scientific knowledge and technologies to the design of buildings.

## Earthquake Engineering

Earthquake engineering is a sub-discipline of civil engineering that influences the life and property of all human beings. It is this science that has provided an in-depth knowledge of earthquakes, useful guidelines to minimize the damage due to earthquakes.

Earthquake engineering is the science of the performance of buildings and structures when subjected to seismic loading. It also assists analysing the interaction between civil infrastructure and the ground, including the consequences of earthquakes on structures. One of the most important aims of earthquake engineering is the proper design and construction of buildings in accordance with building codes, so as to minimize damage due to earthquakes. It is the earthquake engineer who ensures proper design of buildings so they will resist damage due to earthquakes, but at the same time not be unnecessarily expensive.

## Seismic Vibration Control Technologies

The purpose of these technologies is to minimize the seismic effects on buildings and other infrastructure by the use of seismic control devices. When seismic waves start penetrating the base of the buildings from the ground level, the flow density of their energy reduces due to reflections and other reasons. However, the remaining waves possess significant potential for damage when they reach the superstructure.

Vibration control devices assist in the reduction of the damaging effects, and enhance the seismic performance characteristics of the building. When the seismic waves penetrate a superstructure, these are dissipated by the use of dampers, or dispersed in a wide range of frequencies. Mass dampers are also employed to absorb the resonant wave frequencies of seismic waves, thus reducing the damaging effects. Seismic isolation techniques are sometimes used to partly suppress the flow of seismic energy into the superstructure by the insertion of pads into or beneath the load bearing elements in the base of the structure. Thus, the structure is protected from the damaging consequences of an earthquake by decoupling the structure from the shaking ground.

## Façade Engineering

*"Facade Engineering is the art of resolving aesthetic, environmental and structural issues to achieve the enclosure of habitable space."*

With the ever increasing complexity and performance demands on buildings, there is a need to accurately validate the design and communicate the client and designers' intentions to those who have a responsibility for executing the façades of modern buildings.

Façade Engineers are proactive members of the design team and are able to help deliver best value to projects by being fully involved in the early iterative strategic stages.

The façade of a modern building is one of the most expensive and important elements of building construction and can represent up to 35% of construction costs. Façades represent the manifestation of the Architect's expectation of the external appearance of a building. They are the primary environmental modifier and are the elements that have the greatest level of historic failure.

With the implementation of new and more onerous energy requirements, the design of façades must be part of a holistic strategy. In particular science needs to be carefully considered to deal with aspects such as façade performance and energy conservation. There now needs to be consideration of issues such as predictive thermal modelling techniques, testing, lighting and acoustic engineering, added to existing issues of durability, sustainability and structural engineering to drive best value for clients. The blurring of the boundaries is recognised in the fact that the energy conservation and statutory compliances are now a matter of balance between the disciplines.

As the nature of building façades takes on ever greater importance, so proportionally does the level of technology involved. The pressure for change is probably greater than for any other element of a building.

## Fire Engineering

Fire Engineering is the application of scientific and engineering principles, rules [Codes], and expert judgement, based on an understanding of the phenomena and effects of fire and of the reaction and behaviour of people to fire, to protect people, property and the environment from the destructive effects of fire.

These objectives will be achieved by a variety of means including such activities as:

- the assessment of the hazards and risks of fire and its effects.
- the mitigation of potential fire damage by proper design, construction, arrangement, and use of buildings, materials, structures, industrial processes, transportation systems and similar.
- the appropriate level of evaluation for the optimum preventive and protective measures necessary to limit the consequences of fire.
- the design, installation, maintenance and/or development of fire detection, fire suppression, fire control and fire related communication systems and equipment.
- the direction and control of appropriate equipment and manpower in the strategy and function of fire fighting and rescue operations.
- Post-fire investigation and analysis, evaluation and feedback.

The discipline of fire engineering includes, but is not exclusive to:

- Fire detection: fire alarm systems and brigade call systems.
- Active fire protection: fire suppression systems.

- Passive fire protection: fire and smoke barriers, space separation.

- Smoke control and management.

- Escape facilities: emergency exits, fire lifts, etc.

- Building design, layout, and space planning.

- Fire prevention programs.

- Fire dynamics and fire modeling.

- Human behaviour during fire events.

- Risk analysis, including economic factors.

- Wildfire management.

A fire engineer, by education, training and experience understands:

- The nature and characteristics of fire and the mechanisms of fire.

- The spread and the control of fire and the associated products of combustion.

- How fires originate.

- The spread within and outside buildings/structures.

- How fire can be detected, controlled, and/or extinguished.

- Is able to anticipate the behaviour of materials, structures, machines, apparatus, and processes as related to the protection of life, property and the environment from fire.

- Has an understanding of the interactions and integration of fire safety systems in buildings, industrial structures and similar facilities.

- Is able to make use of all of the above and any other required knowledge to undertake the practice of fire engineering.

## Roof Engineering

Roof engineering is a speciality within civil engineering which deals with structure of roofs. Roof is the upper most portion of the building which protects the building from rain, wind and sun. Various types of roofs used may be divided broadly into three types:

1. Flat roofs

2. Pitched roofs

3. Shells and folded plates.

Flat roofs are used in plains where rainfall is less and climate is moderate. Pitched roofs are preferred wherever rainfall is more. Shells and folded plate roofs are used to cover largecolumn free areas required for auditoriums, factories etc. Brief description of these roofs is presented below:

1. Flat Roofs: These roofs are nearly flat. However slight slope (not more than 10°) is given to drain out the rain water. All types of upper storey floors can serve as flat roofs. Many times top of these roofs are treated with water proofing materials-like mixing water proofing chemicals in concrete, providing coba concrete. With advent of reliable water proofing techniques such roofs are constructed even in areas with heavy rain fall.

The advantages of flat roofs are:

(a) The roof can be used as a terrace for playing and celebrating functions.

(b) At any latter stage the roof can be converted as a floor by adding another storey.

(c) They can suit to any shape of the building.

(d) Over-head water tanks and other services can be located easily.

(e) They can be made fire proof easily compared to pitched roof.

The disadvantages of flat roofs are:

(a) They cannot cover large column free areas.

(b) Leakage problem may occur at latter date also due to development of cracks. Once leakage problem starts, it needs costly treatments.

(c) The dead weight of flat roofs is more.

(d) In places of snow fall flat roofs are to be avoided to reduce snow load.

(e) The initial cost of construction is more.

(f) Speed of construction of flat roofs is less.

Types of Flat Roofs: All the types listed for upper floors can be used as flat roofs.

2. Pitched Roofs: In the areas of heavy rain falls and snow fall sloping roof are used. The slope of roof shall be more than 10°. They may have slopes as much as 45° to 60° also. The sloped roofs are known as pitched roofs. The sloping roofs are preferred in large spanned structures like workshops, factory buildings and ware houses. In all these roofs covering sheets like A.C. sheet, G.I. sheets, tiles, slates etc. are supported on suitable structures. The pitched roofs are classified into:

(a) Single roofs.

(b) Double or purlin roofs.

(c) Trussed roofs.

(a) Single Roof: If the span of roof is less than 5 m the following types of single roofs are used.

(i) Lean to roofs.

(ii) Coupled roofs.

(iii)  Coupled-close roof.

(iv)  Collar beam roof.

In all these roofs rafters placed at 600 mm to 800 mm spacing are main members taking load of the roof. Battens run over the rafters to support tiles. Figure shows various types of single roofs.

(b) Double or Purlin Roofs: If span exceeds, the cost of rafters increase and single roof becomes uneconomical. For spans more than 5 m double purlin roofs are preferred. The intermediate support is given to rafters by purlins supported over collar beams. Figure shows a typical double or purlin roof.

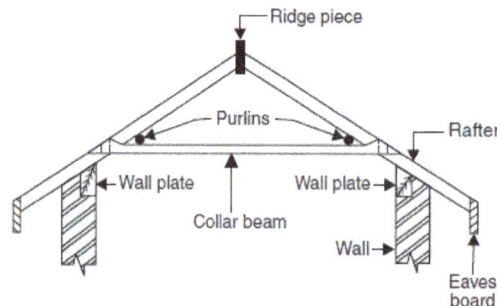

Figure: Double or purlins roofs

(c) Trussed Roof: If span is more, frame works of slender members are used to support sloping roofs.

These frames are known as trusses. A number of trusses may be placed lengthwise to get wall free longer halls. Purlins are provided over the trusses which in turn support roof sheets. For spans up to 9 m wooden trusses may be used but for larger spans steel trusses are a must. In case of wooden trusses suitable carpentry joints are made to connect various members at a joint. Bolts and straps are also used. In case of steel trusses joints are made using gusset plates and by providing bolts or rivets or welding. Depending upon the span, trusses of different shapes are used. End of trusses are supported on walls or on column. Figure shows different shapes of trusses used. Figure shows a typical wooden truss details and figure shows the details of a typical steel truss.

Shells and folded plate roofs: Shell roof may be defined as a curved surface, the thickness of which is small compared to the other dimensions. In these roofs lot of load is transferred by membrane compression instead of by bending as in the case of conventional slab and beam constructions. Caves are having natural shell roofs. An examination of places of worships built in India, Europe and Islamic nations show that shell structures were in usage for the last 800 to 1000 years. However the shells of middle ages were massive masonry structures but nowadays thin R.C.C. shell roofs are built to cover large column free areas. Figure shows commonly used shell roofs.

## Advantages and Disadvantages of Shell Roofs

Advantages of shell roofs are:

(a)  Good from aesthetic point of view.

(b) Material consumption is quite less.

(c) Form work can be removed early.

(d) Large column free areas can be covered.

Disadvantages are:

(a) Top surface is curved and hence advantage of terrace is lost.

(b) Form work is costly.

Folded plate roofs may be looked as slab with a number of folds. These roofs are also known as hipped plates, prismatic shells and faltwerke. In these structures also bending is reduced and lot of load gets transferred as membrane compression. However folded plates are not so efficient as shells.

Figure shows typical folded plate roofs.

## Advantages and Disadvantages of Folded Plate Roofs Over Shell Roofs

Advantages are:

(a) Form work required is relatively simpler.

(b) Movable form work can be employed.

(c) Design involves simpler calculations.

Disadvantages are:

(a) Folded plate consume more material than shells.

(b) Form work can be removed after 7 days while in case of shells it can be little earlier.

## Roof Coverings for Pitched Roofs

Various types of covering materials are available for pitched roofs and their selection depends upon the climatic conditions, fabrication facility, availability of materials and affordability of the owner. Commonly used pitched roof covering materials are:

(a) Thatch

(b) Shingle

(c) Tiles

(d) Slates

(e) Asbestos cement (A.C.) sheets

(f) Galvanised iron (G.I.) sheets

(a) Thatch Covering: These coverings are provided for small spans, mainly for residential buildings in villages. Thatch is a roof covering of straw, reeds or similar materials. The thatch is well-soaked in water or fire resisting solution and packed bundles are laid with their butt ends pointing towards eves. Thickness varies from 150 mm to 300 mm. They are tied with ropes or twines to supporting structures. The supporting structure consists of round bamboo rafters spaced at 200 mm to 300 mm over which split bamboos laid at right angles at close spacing. It is claimed that reed thatch can last 50 to 60 years while straw thatch may last for 20–25 years.

The advantage of thatch roof is they are cheap and do not need skilled workers to build them. The disadvantages are they are very poor fire resistant and harbour rats and other insects.

(b) Shingles: Wood shingles are nothing but the split or sawn thin pieces of wood. Their size varies from 300 mm to 400 mm and length from 60 mm to 250 mm. Their thickness varies from 10 mm at one end to 3 mm at the other end. They are nailed to supporting structures. They are commonly used in hilly areas for low cost housing. They have very poor fire and termite resistance.

(c) Tiles: Various clay tiles are manufactured in different localities. They serve as good covering materials. Tiles are supported over battens which are in turn supported by rafters/trusses etc. Allahabad tiles, Mangalore tiles are excellent inter-locking tiles. They give good appearance also.

(d) Slates: A slate is a sedimentary rock. Its colour is gray. It can be easily split into thin sheets. Slates of size 450 mm to 600 mm wide, 300 mm long and 4 to 8 mm thick are used as covering materials of pitched roofs in the areas where slate quarries are nearby. A good slate is hard, tough, durable. They are having rough texture and they give ringing bell like sound when struck. They do not absorb water.

(e) A.C. Sheets: Asbestos cement is a material which consists of 15 per cent of asbestos fibres evenly distributed and pressed with cement. They are manufactured in sufficiently large size.

The width of a A.C. sheet varies from 1.0 to 1.2 m and length from 1.75 to 3.0 m. To get sufficient strength with thin sections they are manufactured with corrugation or with traffords [Fig. 8.20]. They are fixed to the steel purlins using J-bolts. The roofing is quite economical, waterproof. However not very good thermal resistant. They are commonly used as covering materials in ware houses, godowns or for larger halls. In auditorium etc., if these sheets are used, false ceilings are provided to get good thermal resistance.

(f) G.I. Sheets: Galvanised iron corrugated sheets are manufactured in the sizes 1.0 to 1.2 m wide and 1.65 m length. Galvanisation of iron makes them rust proof. They are fixed to steel purlins using J-bolts and washers. They are durable, fire proof, light in weight and need no maintenance. They are commonly used as covering materials for ware houses, godown, sheds etc.

## Wind Engineering

Wind engineering is a crucial part of many construction projects. At heights of 400 m and above, the wind flow is similar to that experienced by an airplane. The aim is to reduce the building's movement to a level that its occupants are unaware of, and the challenge increases in proportion to the height of the building. This is particularly true of residential buildings, where the criteria are much more stringent than for commercial buildings.

Our solutions include adjusting the shape of a building to make it more aerodynamic, such as introducing openings to allow the wind to pass through, or adding curves at critical locations along the facade to minimize the "vortex shedding" response which causes high acceleration.

The nature and magnitude of wind forces on structures are complex functions of shielding, stagnation, vortex shedding, and other phenomena. Building codes provide approximate relationships between wind speed and wind pressure on building and structure components that are intended to be used for design. More precise mapping of actual wind pressures requires wind-tunnel testing, or in some circumstances, computer modeling of the wind flow. However, experience shows that wind pressures are typically highest at architectural discontinuities high on the structure, such as roof eaves, dormers, or parapet corners, and wind damage often starts at these locations.

# Earthquake-resistant Structures

Earthquake-resistant structure is a Building designed to prevent total collapse, preserve life, and minimize damage in case of an earthquake or tremor. Earthquakes exert lateral as well as vertical forces, and a structure's response to their random, often sudden motions is a complex task that is just beginning to be understood. Earthquake-resistant structures absorb and dissipate seismically induced motion through a combination of means: damping decreases the amplitude of oscillations of a vibrating structure, while ductile materials (e.g., steel) can withstand considerable inelastic deformation. If a skyscraper has too flexible a structure, then tremendous swaying in its upper floors can develop during an earthquake. Care must be taken to provide built-in tolerance for some structural damage, resist lateral loading through stiffeners (diagonal sway bracing), and allow areas of the building to move somewhat independently.

## Earthquake Resistant Buildings by Earthquake Engineering

To be earthquake proof, buildings, structures and their foundations need to be built to be resistant to sideways loads. The lighter the building is, the less the loads. This is particularly so when the weight is higher up. Where possible the roof should be of light-weight material. If there are floors and walls and partitions, the lighter these are the better, too.

If the sideways resistance is to be obtained from walls, these walls must go equally in both directions. They must be strong enough to take the loads. They must be tied in to any framing, and reinforced to take load in their weakest direction. They must not fall apart and must remain in place after the worst shock waves so as to retain strength for the after shocks.

If the sideways resistance comes from diagonal bracing then it must also go equally all round in both directions. Where possible, it should be strong enough to accept load in tension as well as compression: the bolted or welded connections should resist more tension than the ultimate tension value of the brace (or well more than the design load) and it should not buckle with loads well above the design load.

And the loads have got to go down to ground in a robust way. If the sideways load is to be resisted with moment resisting framing then great care has to be taken to ensure that the joints are stronger than the beams, and that the beams will fail before the columns, and that the columns cannot fail by spelling if in concrete. Again the rigid framing should go all around, and in both directions.

If the building earthquake resistance is to come from moment resisting frames, then special care should be taken with the foundation-to-first floor level. If the requirement is to have a taller clear height, and to have open holes in the walls, then the columns at this level may have to be much stronger than at higher levels; and the beams at the first floor, and the columns from ground to second floor, have to be able to resist the turning loads these columns deliver to the frame.

Alternatively, and preferably, the columns can be given continuity at the feet. This can be done with 'fixed feet' with many bolts into large foundations, or by having a grillage of steel beams at the foundation level able to resist the column moments. Such steel grillage can also keep the foundations in place.

If the beams in the frame can bend and yield a little at their highest stressed points, without losing resistance, while the joints and the columns remain full strength, then a curious thing happens: the resonant frequency of the whole frame changes. If the building was vibrating in time with shock waves, this vibration will tend to be damped out.

This phenomenon is known as 'plastic hinging' and is easily demonstrated in steel beams, though a similar thing can happen with reinforced concrete beams as long as spelling is avoided.

All floors have to be connected to the framing in a robust and resilient way. They should never be able to shake loose and fall. Again all floors should be as light as possible. They should go all round each column and fix to every supporting beam or wall, in a way that cannot be shaken off.

One way of reducing the vulnerability of big buildings is to isolate them from the floor using bearings or dampers, but this is a difficult and expensive process not suitable for low and medium rise buildings and low cost buildings.

## Designing of Earthquake Resistant Structures

When designing earthquake safe structures the first consideration is to make the highest bit, the roof, as light as possible. This is best done with profiled steel cladding on light gauge steel Zed purlins. This can also have double skin with spacers and insulation. It can have a roof slope between 3 and 15 degrees. If it is required to have a 'flat' roof, this could be made with a galvanised steel decking and solid insulation boards, and topped with a special membrane. Even a 'flat' roof should have a slope of about 2 degrees.

If it is required to have a 'flat' concrete roof, then the best solution is to have steel joists at about 2m, 6", centres, and over these to have composite style roof decking. Then an RC slab can be poured over the roof, with no propping; the slab will only be say 110mm, 4 1/2", and will weigh only about 180 kg/sqm. Such a slab will be completely bonded to the frame and will not be able to slip off, or collapse.

If the building or structure is a normal single storey, then any normal portal frame or other steel framed building, if the design and construction is competently done, will be resistant to Earthquake loads. If it is to have 2 or more stories, more needs to be done to ensure its survival in an earthquake. As with the roof, the floors should be made as light as possible. The first way to do this is to use traditional timber joists and timber or chipboard or plywood flooring. If this is done it is vital that the timber joists are firmly through bolted on the frames to avoid them slipping or being torn off. The frame needs them for stability and the floor must never fall down.

A better alternative is to substitute light gauge steel Zeds for the timber joists. These can span further and are easier to bolt firmly to the framework. Then, floor-boards or tongue-and-groove chipboard can easily be screwed to the Zeds. However in Hotels, Apartment buildings, Offices and the like, concrete floors may be needed. In such cases we should reduce the spans to the spanning capacity of composite decking flooring, and pour reinforced concrete slabs onto our decking. The decking is fixed to the joists, the joists into the main beams, the main beams into the columns and the concrete is poured around all the columns. There is simply no way that such floors can fall off the frame.

## Proof Building Diagram

SPECIAL MOMENT RESISTING FRAMES TO PREVENT EARTHQUAKE DAMAGE

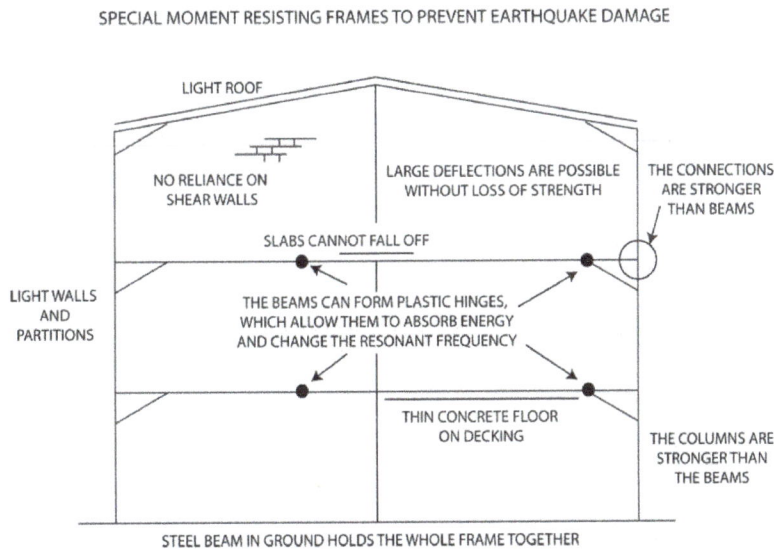

LIGHT ROOF

NO RELIANCE ON SHEAR WALLS

LARGE DEFLECTIONS ARE POSSIBLE WITHOUT LOSS OF STRENGTH

THE CONNECTIONS ARE STRONGER THAN BEAMS

SLABS CANNOT FALL OFF

LIGHT WALLS AND PARTITIONS

THE BEAMS CAN FORM PLASTIC HINGES, WHICH ALLOW THEM TO ABSORB ENERGY AND CHANGE THE RESONANT FREQUENCY

THIN CONCRETE FLOOR ON DECKING

THE COLUMNS ARE STRONGER THAN THE BEAMS

STEEL BEAM IN GROUND HOLDS THE WHOLE FRAME TOGETHER

Once the floors are robustly fitted to the frames, the frames themselves must be correctly designed. Please look at the diagram above.

Start at the bottom. The frame should not be built on simple pinned feet at ground level. Outside earthquake zones it is normal to build a 'nominally pinned footing' under each column. This actually gives some fixity to the base as well as horizontal and vertical support. But in an earthquake, this footing may be moving and rotating, so rather than provide a bit of fixity, it can push to left or right, or up and down, and rotate the column base, helping the building to collapse prematurely.

Any pinned footing may actually be moving differently from other footings on the same building, and so not even be giving horizontal or vertical support, but actually helping to tear the building apart. So to earthquake proof the building REIDsteel would start with steel ground beams joining the feet together, and these should have moment resistance to prevent the bottoms of the columns from rotating.

These ground beams may well go outside the line of the building, thus effectively reducing the height-to-width ratio as well, helping to reduce total over-turning. This ground beam may be built on pads or piles or rafts as appropriate. On loose soils, the bearing pressure should be very conservatively chosen, to minimise effect of liquefaction.

Among the most important advanced techniques of earthquake resistant design and construction are:

- Base Isolation

- Energy Dissipation Devices

## Base Isolation Method of Earthquake Resistant Design

A base isolated structure is supported by a series of bearing pads which are placed between the building and the building's foundation. A variety of different types of base isolation bearing pads have now been developed.

The bearing is very stiff and strong in the vertical direction, but flexible in the horizontal direction.

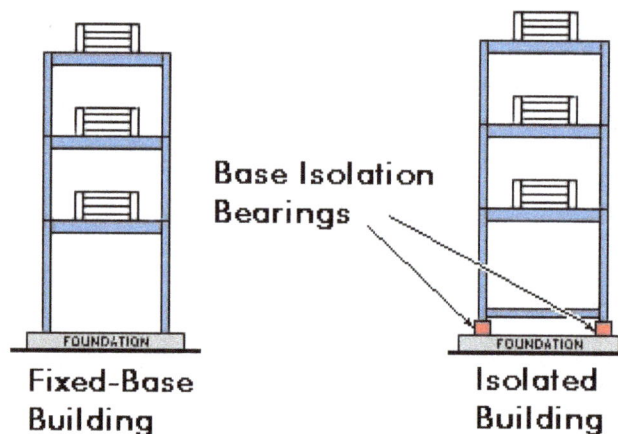

Figure: Base-Isolated and Fixed-Base Buildings

## Earthquake Generated Forces

To get a basic idea of how base isolation works, examine Figure. This shows an earthquake acting on both a base isolated building and a conventional, fixed-base, and building. As a result of an earthquake, the ground beneath each building begins to move. In Figure, it is shown moving to the left.

Each building responds with movement which tends toward the right. The building undergoes displacement towards the right. The building's displacement in the direction opposite the ground motion is actually due to inertia. The inertial forces acting on a building are the most important of all those generated during an earthquake.

It is important to know that the inertial forces which the building undergoes are proportional to the building's acceleration during ground motion.

It is also important to realize that buildings don't actually shift in only one direction. Because of the complex nature of earthquake ground motion, the building actually tends to vibrate back and forth in varying directions.

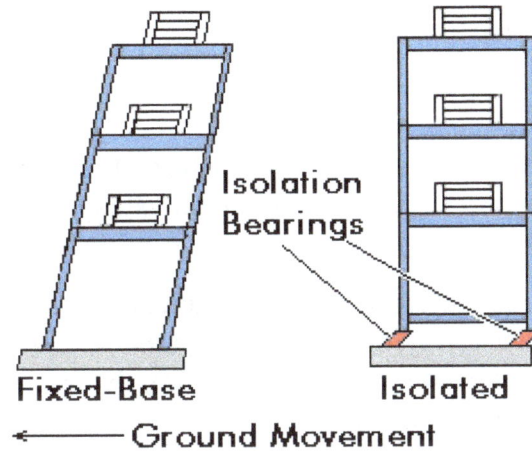

Figure: Base-Isolated, Fixed-Base Buildings

## Deformation and Damages to Structures

In addition to displacing toward the right, the un-isolated building is also shown to be changing its shape-from a rectangle to a parallelogram. It is deforming. The primary cause of earthquake damage to buildings is the deformation which the building undergoes as a result of the inertial forces acting upon it.

## Response of Base Isolated Building

By contrast, even though it too is displacing, the base-isolated building retains its original, rectangular shape. It is the lead-rubber bearings supporting the building that are deformed.

The base-isolated building itself escapes the deformation and damage, which implies that the inertial forces acting on the base-isolated building have been reduced.

Experiments and observations of base-isolated buildings in earthquakes have been shown to reduce building accelerations to as little as 1/4 of the acceleration of comparable fixed-base buildings, which each building undergoes as a percentage of gravity.

As we noted above, inertial forces increase, and decrease, proportionally as acceleration increases or decreases.

Acceleration is decreased because the base isolation system lengthens a building's period of vibration, the time it takes for the building to rock back and forth and then back again. And in general, structures with longer periods of vibration tend to reduce acceleration, while those with shorter periods tend to increase or amplify acceleration.

Finally, since they are highly elastic, the rubber isolation bearings don't suffer any damage. But the lead plug in the middle of our example bearing experiences the same deformation as the rubber. However, it generates heat.

In other words, the lead plug reduces, or dissipates, the energy of motion, i.e., kinetic energy–by converting that energy into heat. And by reducing the energy entering the building, it helps to slow and eventually stop the building's vibrations sooner than would otherwise be the case, in other words, it damps the building's vibrations.

## Energy Dissipation Devices

The second of the major new techniques for improving the earthquake resistance of buildings also relies upon damping and energy dissipation, but it greatly extends the damping and energy dissipation provided by lead-rubber bearings.

As we've said, a certain amount of vibration energy is transferred to the building by earthquake ground motion. Buildings themselves do possess an inherent ability to dissipate, or damp, this energy. However, the capacity of buildings to dissipate energy before they begin to suffer deformation and damage is quite limited.

The building will dissipate energy either by undergoing large scale movement or sustaining increased internal strains in elements such as the building's columns and beams. Both of these eventually result in varying degrees of damage.

So, by equipping a building with additional devices which have high damping capacity, we can greatly decrease the seismic energy entering the building, and thus decrease building damage.

Accordingly, a wide range of energy dissipation devices have been developed and are now being installed in real buildings. Energy dissipation devices are also often called damping devices. The large number of damping devices that have been developed can be grouped into three broad categories:

- Friction Dampers: these utilize frictional forces to dissipate energy
- Metallic Dampers : utilize the deformation of metal elements within the damper
- Viscoelastic Dampers : utilize the controlled shearing of solids
- Viscous Dampers: utilized the forced movement (orificing) of fluids within the damper

## Fluid Viscous Dampers

General principles of damping devices are illustrated through Fluid Viscous damper. Following section, describes the basic characteristics of fluid viscous dampers, the process of developing and testing them, and the installation of fluid viscous dampers in an actual building to make it more earthquake resistant.

## Damping Devices and Bracing Systems

Damping devices are usually installed as part of bracing systems. Figure 3 shows one type of damper-brace arrangement, with one end attached to a column and one end attached to a floor beam. Primarily, this arrangement provides the column with additional support.

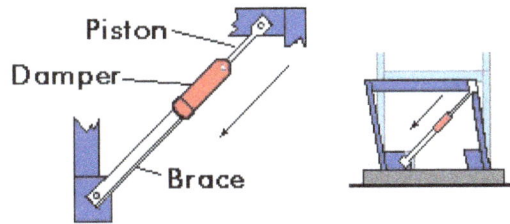

Figure: Damping Device Installed with Brace

Most earthquake ground motion is in a horizontal direction; so, it is a building's columns which normally undergo the most displacement relative to the motion of the ground. Figure 3 also shows the damping device installed as part of the bracing system and gives some idea of its action.

## References

- Earthquake-engineering-goals-technology-and-research-methods-41861: brighthubengineering.com, Retrieved 21 May 2018

- What-is-facade-engineering, facade-engineering: wintech-group.com, Retrieved 31 March 2018

- What-is-a-fire-engineer: firesafe.org.uk, Retrieved 28 June 2018

- Wind-engineering-and-modelling: wsp.com, Retrieved 14 July 2018

- Earthquake-resistant-structure, technology: britannica.com, Retrieved 28 March 2018

- Earthquake-resistant-techniques-5607: theconstructor.org, Retrieved 18 March 2018

# 2

# Structural Components

Every engineering structure is made up of many structural elements, such as columns, arches, beams, plates, shells, etc. These elements are used to build complex structural systems. The aim of this chapter is to explore the varied aspects of structural elements and systems, such as rigid frame, braced frame, space frame, eyebar, tensile structure, etc.

## Structural Element

A structural element is related to their parent topology by nature. Structural elements are needed to receive the tension loads exerted on the cloth either by stabilization prestressing or external actions, such as wind or snow. The type used depends on the loads and the position of the element, bearing in mind that it may serve as a high or a low point in the membrane.

Any structure is essentially made up of only a small number of different types of elements:

- Columns
- Beams
- Plates
- Arches
- Shells
- Catenaries

Many of these elements can be classified according to form (straight, plane/curve) and dimensionality (one-dimensional/two-dimensional):

| | One-dimensional | | Two-dimensional | |
|---|---|---|---|---|
| | **straight** | **curve** | **plane** | **curve** |
| (predominantly) bending | beam | continuous arch | plate, concrete slab | lamina, dome |
| (predominant) tensile stress | rope | Catenary | shell | |
| (predominant) compression | pier, column | | Load-bearing wall, shell | |

### Columns

Columns are elements that carry only axial force—either tension or compression—or both axial force and bending (which is technically called a beam-column but practically, just a column). The design of a column must check the axial capacity of the element, and the buckling capacity.

The buckling capacity is the capacity of the element to withstand the propensity to buckle. Its capacity depends upon its geometry, material, and the effective length of the column, which depends upon the restraint conditions at the top and bottom of the column. The effective length is $K*l$ where $l$ is the real length of the column.

The capacity of a column to carry axial load depends on the degree of bending it is subjected to, and vice versa. This is represented on an interaction chart and is a complex non-linear relationship.

## Beams

A beam may be:

- Cantilevered (supported at one end only with a fixed connection)

- Simply supported (supported vertically at each end but able to rotate at the supports)

- Continuous (supported by three or more supports)

- A combination of the above (ex. supported at one end and in the middle)

Beams are elements which carry pure bending only. Bending causes one section of a beam (divided along its length) to go into compression and the other section into tension. The compression section must be designed to resist buckling and crushing, while the tension section must be able to adequately resist the tension.

## Struts and Ties

*Little Belt*: a truss bridge in Denmark

The McDonnell Planetarium by Gyo Obata in St Louis, Missouri, U.S., a concrete shell structure

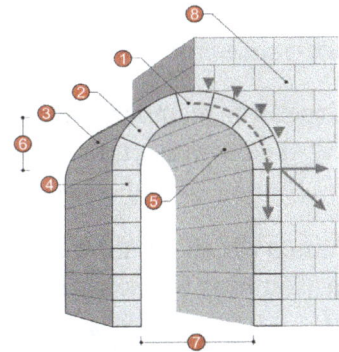

A masonry arch 1. Keystone 2. Voussoir 3. Extrados 4. Impost 5. Intrados 6. Rise 7. Clear span 8. Abutment

A truss is a structure comprising two types of structural element, i.e. struts and ties. A strut is a relatively lightweight column and a tie is a slender element designed to withstand tension forces. In a pin-jointed truss (where all joints are essentially hinges), the individual elements of a truss theoretically carry only axial load. From experiments it can be shown that even trusses with rigid joints will behave as though the joints are pinned.

Trusses are usually utilized to span large distances, where it would be uneconomical and unattractive to use solid beams.

## Plates

Plates carry bending in two directions. A concrete flat slab is an example of a plate. Plates are understood by using continuum mechanics, but due to the complexity involved they are most often designed using a codified empirical approach, or computer analysis.

They can also be designed with yield line theory, where an assumed collapse mechanism is analysed to give an upper bound on the collapse load. This is rarely used in practice.

## Shells

Shells derive their strength from their form, and carry forces in compression in two directions. A dome is an example of a shell. They can be designed by making a hanging-chain model, which will act as a catenary in pure tension, and inverting the form to achieve pure compression.

## Arches

Arches carry forces in compression in one direction only, which is why it is appropriate to build arches out of masonry. They are designed by ensuring that the line of thrust of the force remains within the depth of the arch.

## Catenaries

Catenaries derive their strength from their form, and carry transverse forces in pure tension by deflecting (just as a tightrope will sag when someone walks on it). They are almost always cable or fabric structures. A fabric structure acts as a catenary in two directions.

## Compression Member

A structural member loaded axially in compression is generally called a compression member. Vertical compression members in buildings are called columns, posts or stanchions. A compression member in roof trusses is called struts and in a crane is called a boom.

Columns which are short are subjected to crushing and behave like members under pure compression. Columns which are long tend to buckle out of the plane of the load axis.

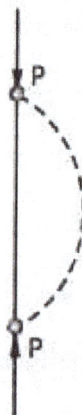

## Theory of Columns

Euler's formula for critical load for a pin-ended column subjected to axial load is

$$P_{cr} = \frac{\pi^2 EI}{L^2}$$

Where, L = length of column between the hinged ends,

E = modulus of elasticity, and

I = moment of inertia of the column section.

The column will become unserviceable if the loads are larger than $P_{cr}$ In the Euler equation, it is assumed that stress is proportional to strain, therefore,

Critical Stress =

$$\frac{P_{cr}}{A} = \frac{\pi^2 EI}{AL^2} = \frac{\pi^2 EI}{\left(\dfrac{L}{r}\right)^2} = \frac{\pi^2 E}{\lambda^2}$$

Where, A= area of cross-section, and

r = radius of gyration about the bending axis

$\lambda$ = slenderness ratio

## Various end Conditions

Columns with length L and effective length $l$ are shown in figure below:

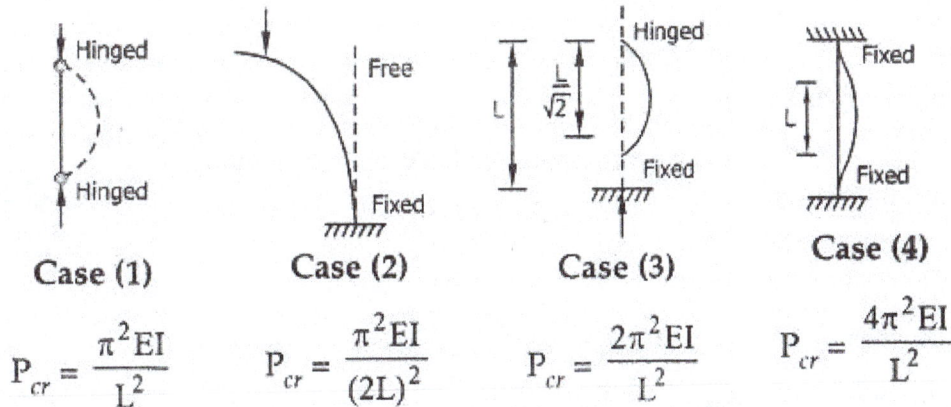

| Case (1) | Case (2) | Case (3) | Case (4) |
|---|---|---|---|
| $P_{cr} = \dfrac{\pi^2 EI}{L^2}$ | $P_{cr} = \dfrac{\pi^2 EI}{(2L)^2}$ | $P_{cr} = \dfrac{2\pi^2 EI}{L^2}$ | $P_{cr} = \dfrac{4\pi^2 EI}{L^2}$ |

## Strength of an Axially Loaded Compression Members

Maximum axial compression load permitted on a compression member,

$$P = \sigma_{ac} \times A$$

Where, P = axial compressive load (N),

$\sigma_{ac}$ = permissible stress in axial compression (MPa)

A = effective cross-sectional area of the member (mm²)

It stipulates that the direct stress on the cross-sectional area of axially loaded compression members should not exceed 0.6 $f_y$ nor the permissible stress calculated using Merchant – Rankine formula.

Permissible stress in axial compression (MPa):

$$\sigma_{ac} = 0.6 \times \frac{f_{cc} \times f_y}{\left[f^n{}_{cc} \times f^n{}_y\right]^{1/n}}$$

Where $f_y$ = yield stress of steel in MPa

$f_{cc}$ = elastic critical stress in compression = $\dfrac{\pi^2 E}{\lambda^2}$

$\lambda = \dfrac{l}{r}$ = slenderness ratio of the member

Where, $l$ = effective length of the member

r = appropriate radius of gyration of the member

E = modulus of elasticity = 200000 MPa, and

n = a factor assumed as 1.4

## Effective Length of Compression Member

Table below gives the values of effective length recommended by the Indian Standard, IS 800. The actual length L of the compression member should be taken as the length from centre-to-centre of intersection of supporting members or the cantilevered length in the case of free standing struts.

Table: Equivalent length for various end conditions

|   | Type | Effective length of member l |
|---|------|------------------------------|
| 1 | Effectively held in position and restrained in direction at both ends. | 0.67 L |
| 2 | Effectively held in position at both ends restrained in direction at one end. | 0.85 L |
| 3 | Effectively held in position at both ends but not restrained in direction. | L |
| 4 | Effectively held in position and restrained in direction at one end and at the other end effectively restrained in direction but not held in position. | L |
| 5 | Effectively held in position and restrained in direction at one end and the other end partially restrained in direction but not held in position. | 1.5 L |
| 6 | Effectively held in position and restrained in direction at one end but not held in position or restrained in direction at the other end. | 2.0 L |

Note:

1.  L is the unsupported length of compression member.

2.  For battened struts, the effective length should be increased by 10%.

Maximum slenderness ratio:

According to Indian Standard IS 800, the slenderness ratio should not exceed the values given in the table below:

| No. | Type of Member | Slenderness ratio $\lambda = \dfrac{l}{r}$ |
|-----|----------------|--------------------------------------------|
| 1 | A member carrying compressive loads resulting from dead and superimposed loads. | 180 |
| 2 | A member subjected to compressive loads resulting from wind/earthquake forces provided the deformation of such members does not adversely affect the stress in any part of the structure. | 250 |
| 3 | A member normally carrying tension but subjected to reversal of stress due to wind or earthquake forces. | 350 |
| 4 | Tension member (other than pre-tensioned member) | 400 |

## Angle Struts

Single angle discontinuous struts connected by a single rivet or bolt may be designed for axial load only provided the compressive stress does not exceed $0.80\sigma_{ac}$. The value of $\sigma_{ac}$ can be determined on the basis that the effective length l of the strut is from centre-to-centre of inter-section at each end and r is the minimum radius of gyration. In no case, the $\dfrac{l}{r}$ ratio for single angle struts should exceed 180. If a single discontinuous strut is connected by a weld or by two or more rivets or bolts in line along the angle at each end, it may be designed for axial load only provided the compression stress does not exceed $\sigma_{ac}$ arrived at on the basis that l is taken as 0.85 times the length of the strut, centre to centre at each end and r is the minimum radius of gyration.

For double angle struts which are discontinuous, back to back connected to both sides of the gusset or section by not less than two bolts or rivets in line along the angles at each end or by equivalent welding, the load may be regarded as applied axially. The effective length l in the plane of end gusset could be taken between 0.7 and 0.85 times the distance between the intersection depending on the restraint provided, the plane perpendicular to that of the end gusset, the effective length should be taken as equal to the distance between centers of intersections. The calculated average compressive stress should not exceed values of $\sigma_{ac}$ obtained for the appropriate slenderness ratios. The angles should be connected together with tack rivets or welds at intervals along their lengths.

## Compression Members Composed of back-to-back Components

A compression member composed of two angles, channels or tees, back to back, in contact or separated by a small distance should be connected together by riveting, bolting or welding so that the slenderness ratio of each member between the connections is not greater than 40 nor greater than 0.60 times the most unfavorable slenderness ratio of the strut as a whole. In no case, the spacing of tacking rivets in a line exceeds 600mm for such members.

For other types of built-up compression members, where cover plates are used, the pitch of tacking rivets should not exceed 32t or 300mm, whichever is less, where t is the thickness of the thinner outside plate. Where plates are exposed to bad weather conditions, the pitch should not exceed 16 t or 200mm whichever is less.

The rivets, welds and bolts in these connections should be sufficient to carry the shear force and bending moments, if any, specified for battened struts. The diameter of the connecting rivets should not be less than the minimum diameter given in the table below:

| Thickness of member | Minimum diameter of rivets |
| --- | --- |
| Upto 10mm | 16mm |
| Over 10mm upto 16mm | 20mm |
| Over 16mm | 22 mm |

Solid packing or washers should be used for riveting, bolting, where the members are separated back to back.

The end struts should be connected together with not less than two rivets or bolts or their equivalent in welding and there should be not less than two additional connections spaced equidistant in the length of the strut.

A minimum of two rivet or bolts should be used in each connection, one on line of each gauge mark, where the legs of the connected angles or tables of the connected tees are 125mm wide or over, or where the webs of channel are 150mm wide or over.

## Lacings and Battens for Built-up Compression Members

As per Indian Standard, IS 800-1984, the following specifications are used for the design of lacing and batten plates.

In a built-up section, the different components are connected together so that they act as a single column. Lacing is generally preferred in case of eccentric loads. Battening is normally used for axially loaded columns and in sections where the components are not far apart. Flat bars are generally used for lacing. Angles, channels and tubular sections are also used for lacing of very heavily columns. Plates are used for battens.

## Lacings

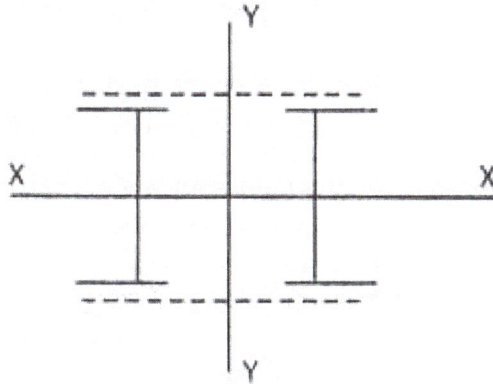

A lacing system should generally conform to the following requirements:

1.  The compression member comprising two main components laced and tied should, where practicable, have a radius of gyration about the axis perpendicular to the plane of lacing not less than the radius of gyration at right angles to that axis.

2.  The lacing system should not be varied throughout the length of the strut as far as practicable.

3.  Cross (except tie plates) should not be provided along the length of the column with lacing system, unless all forces resulting from deformation of column members are calculated and provided for in the lacing and its fastening.

4.  The single-laced systems on opposite sides of the main components should preferably be in the same direction so that one system is the shadow of the other.

5.  Laced compression members should be provided with tie plates at the ends of the lacing system and at points where the lacing system are interrupted. The tie plates should be designed by the same method as followed for battens.

## Guidelines for the Design of Lacing System

1.  The angle of inclination of the lacing with the longitudinal axis of the column should be between 40° to 70°.

2.  The slenderness ratio $\lambda = \dfrac{l_e}{r}$ of the lacing bars should not exceed 145.

3.  The effective length $l_e$ of the lacing bar should be according to the table given below:

| No. | Type of lacing | Effective length, $l_e$ |
|-----|----------------|--------------------------|
| 1 | Single lacing, riveted at ends | Length between inner end rivets on lacing bar. |
| 2 | Double lacing, riveted at ends | 0.7 times the length between end rivets on lacing bars. |
| 3 | Welded lacing | 0.7 times the distance between inner ends or effective lengths of welds at ends. |

1.  For riveted or welded lacing system, $\left(\dfrac{L}{r_{e,min}}\right) \not> 50$ or 0.7 times maximum slenderness ratio of the compression member as a whole, whichever is less.

Here, L = distance between the centers of connections of the lattice bars, and $r_{e,min}$ = the minimum radius of gyration of the components of the compression member.

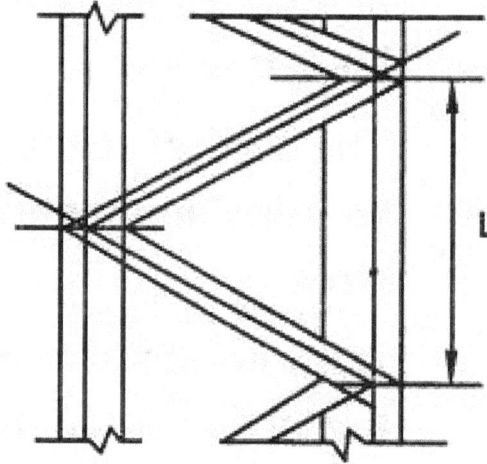

1.  Minimum width of lacing bars in riveted connection should be according to the table given below:

| Nominal rivet diameter (mm) | 22 | 20 | 18 | 16 |
|---|---|---|---|---|
| Width of lacing bars (mm) | 65 | 60 | 55 | 50 |

2.  Minimum thickness of lacing bars:

$$t \not< \frac{l}{40} \text{ , for single lacing;}$$

$$t \not< \frac{l}{60} \text{ , for double lacing;}$$

Where $l$ = length between inner end rivets.

1.  The lacing of compression members should be designed to resist a transverse shear, V=2.5 percent of the axial force in the member. The shear is divided equally among all transverse lacing systems in parallel planes. The lacing system should also be designed to resist additional shear due to bending if the compression member carries bending due to eccentric load, applied end moments, and/or lateral loading.

2.  The riveted connections may be made in two ways, as shown in the figure (a) and (b).

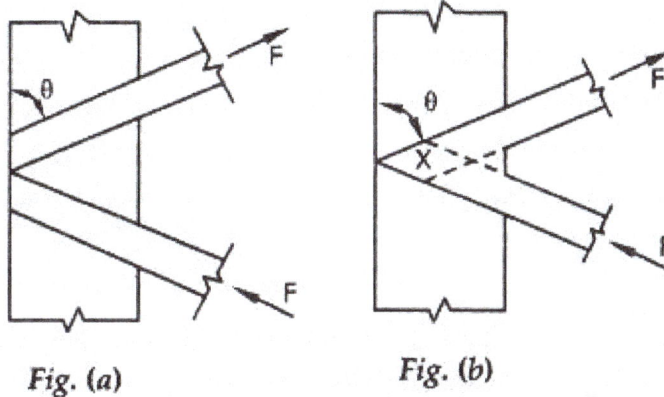

Fig. (a)                                    Fig. (b)

## Welded Connection

Lap Joint: Overlap should not be less than ¼ times thickness of bar or member, whichever is less.

Butt Joint: Full penetration butt weld or fillet weld on each side, lacing bar should be placed opposite to flange or stiffening component of main member.

## Battens

Compression members composed of two main components battened should preferably have these components of the same cross-section and symmetrically disposed about their X – X axis.

The battens should be placed opposite to each-other at each end of the member and at points where the member is stayed in its length, and should as far as practicable, be spaced and proportioned uniformly throughout.

The effective length of columns should be increased by 10 percent.

## Design Details of Battens

1. Spacing of batten C, from centre-to-centre of end fastening should be such that the slenderness ratio of the lesser main component, $\left(\dfrac{C}{r_{e,min}}\right) \not> 50$ or 0.7 times the slenderness ratio of the compression member as a whole about X – X axis (parallel to battens) whichever is less.

2. Effective depth of battens, d shall be taken as distance between end rivets or end welds.

$d > \dfrac{3}{4}a$ for intermediate batten.

d > a, for end batten.

d > 2b , for any batten.

where d = effective depth of batten.

a = centroidal distance of members.

b = width of members in the plane of battens.

1.   Thickness of battens, $t > \left( \dfrac{l_b}{50} \right)$

Where, $l_b$ = distance between the innermost connecting line of rivets or welds.

## Design of Battens

Battens should be designed to carry bending moment and shear arising from a transverse shear,

$$V = \frac{2.5}{100} P$$

Where P = total axial load in the compression member.

Transverse shear V is divided equally between the parallel planes of battens. Battens and their connections to main components resist simultaneously a longitudinal shear.

$$V_1 = \frac{V \times C}{N \times S}$$

and , moment $M = \dfrac{V \times C}{2N}$

due to transverse shear V.

where, C = spacing of battens

N= number of parallel planes of battens

S= minimum transverse distance between centroid of rivet group or welding.

The end connections should also be designed to resist the longitudinal shear force $V_1$ and the moment M.

## For Welded Connection

1.   Lap < 4t

2.   Total length of weld at edge of batten < D/2
     a + b+ c < D/2

let t = thickness of batten

Length of continuous weld at each edge of batten < 1/3 of total length required.

Return weld along transverse axis of the column < 4t

Where, t and D are the thickness and overall depth of battens, respectively.

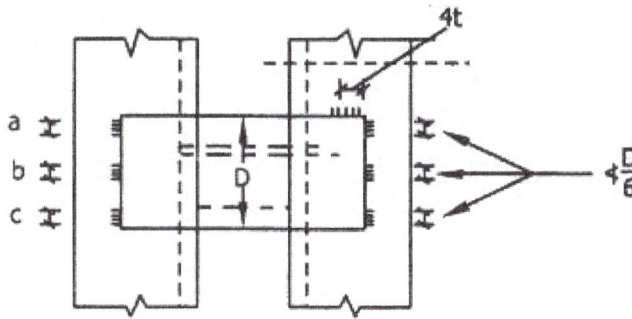

## Beam (Structure)

Any structure member which cross section is much smaller compare to its length and undergoes lateral load, known as beam.

In other words beam is a horizontal bar witch undergoes lateral load or couple which tends to bend it or a horizontal bar undergoes bending stress known as beam.

According to its requirement, different beams use in different conditions like fix beam, cantilever beam etc.

## Types of Beams

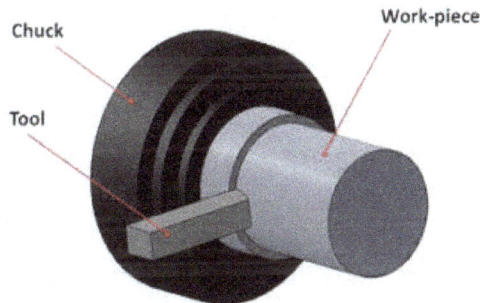

## Beams are Classified

## Simply Supported Beam

As the name implies, simply supported beam is supported at both end. One end of the beam is supported by hinge support and other one by roller support. This support allow to horizontal movement of beam. It beam type undergoes both shear stress and bending moment.

**Simply Supported Beam**

## Continuous Beams

When we talk about types of beams we cannot forgot continuous beam. This beam is similar to simply supported beam except more than two support are used on it. One end of it is supported by hinged support and other one is roller support. One or more supports are use between these beams. It is used in long concrete bridges where length of bridge is too large.

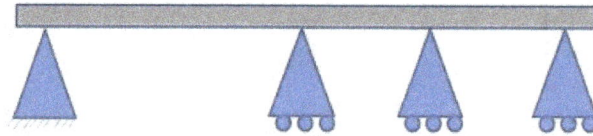

**Continuous Beam**

## Overhanging Beams

Overhanging beam is combination of simply supported beam and cantilever beam. One or both of end overhang of this beam. This beam is supported by roller support between two ends. This type of beam has heritage properties of cantilever and simply supported beam.

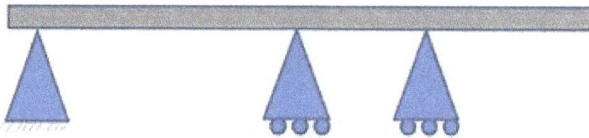

**Overhanging Beam**

## Cantilever Beams

Cantilever beams a structure member of which one end is fixed and other is free. This is one of the famous type of beam use in trusses, bridges and other structure member. This beam carry load over the span which undergoes both shear stress and bending moment.

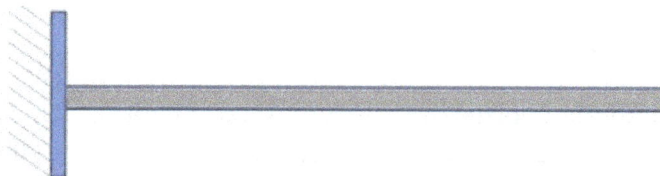

**Cantilever Beam**

## Fixed Beams

This beam is fixed from both ends. It does not allow vertical movement and rotation of the beam. It is only under shear stress and no moment produces in this beams. It is used in trusses, and other structure.

**Fixed Beam**

According to cross section:

A beam may have different cross section. The most common cross section of beam are as follow.

1. I beam:

This beam types have I cross section as shown in figure. It has high resistance of bending.

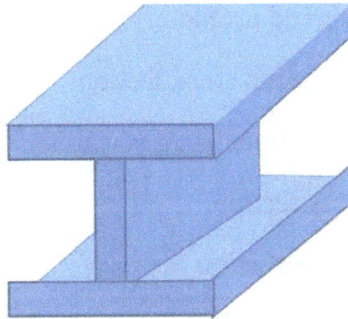

# I Beam

2. T beam:

It has T cross section as shown in figure.

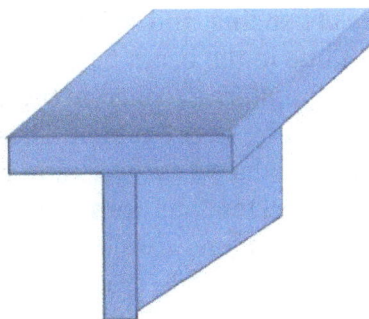

# T Beam

According to equilibrium condition:

1. Statically determinate beam:

A beam is called determinate beam if it can be analyse by the basic equilibrium condition. The support reaction can be found by using basic equilibrium condition. These conditions are Summation of all horizontal forces is zero. Summation of all vertical forces is zero. Summation of all moments is zero. Example: Simply supported beam, Cantilever beam etc.

2. Statically indeterminate beam:

If the beam cannot be analysis by using basic equilibrium condition, known as statically indeterminate beam. The end reaction find out by using basic equilibrium condition with combination of other conditions like strain energy method, virtual work method etc.

Example: Continuous beam, fixed beam.

## According to Geometry

- Straight beam: Beam with straight profile

- Curved beam: Beam with curved profile

- Tapered beam: Beam with tapered cross section

- Based on the shape of cross section:

  1. I-beam: Beam with 'I' cross section

  2. T-beam: Beam with 'T' cross section

  3. C-beam: Beam with 'C' cross section.

Internally, beams experience compressive, tensile and shear stresses as a result of the loads applied to them. Typically, under gravity loads, the original length of the beam is slightly reduced to enclose a smaller radius arc at the top of the beam, resulting in compression, while the same original beam length at the bottom of the beam is slightly stretched to enclose a larger radius arc, and so is under tension. The same original length of the middle of the beam, generally halfway between the top and bottom, is the same as the radial arc of bending, and so it is under neither compression nor tension, and defines the neutral axis (dotted line in the beam figure). Above the supports, the beam is exposed to shear stress. There are some reinforced concrete beams in which the concrete is entirely in compression with tensile forces taken by steel tendons. These beams are known as prestressed concrete beams, and are fabricated to produce a compression more than the expected tension under loading conditions. High strength steel tendons are stretched while the beam is cast over them. Then, when the concrete has cured, the tendons are slowly released and the beam is immediately under eccentric axial loads. This eccentric loading creates an internal moment, and, in turn, increases the moment carrying capacity of the beam. They are commonly used on highway bridges.

The primary tool for structural analysis of beams is the Euler–Bernoulli beam equation. Other mathematical methods for determining the deflection of beams include "method of virtual work" and the "slope deflection method". Engineers are interested in determining deflections because the beam may be in direct contact with a brittle material such as glass. Beam deflections are also minimized for aesthetic reasons. A visibly sagging beam, even if structurally safe, is unsightly and to be avoided. A stiffer beam (high modulus of elasticity and high second moment of area) creates less deflection.

Mathematical methods for determining the beam forces (internal forces of the beam and the forces that are imposed on the beam support) include the "moment distribution method", the force or flexibility method and the direct stiffness method.

## General Shapes

Most beams in reinforced concrete buildings have rectangular cross sections, but a more efficient cross section for a beam is an I or H section which is typically seen in steel construction. Because of the parallel axis theorem and the fact that most of the material is away from the neutral axis, the second moment of area of the beam increases, which in turn increases the stiffness.

An I-beam is only the most efficient shape in one direction of bending: up and down looking at the profile as an I. If the beam is bent side to side, it functions as an H where it is less efficient. The most efficient shape for both directions in 2D is a box (a square shell) however the most efficient shape for bending in any direction is a cylindrical shell or tube. But, for unidirectional bending, the I or wide flange beam is superior.

Efficiency means that for the same cross sectional area (volume of beam per length) subjected to the same loading conditions, the beam deflects less.

Other shapes, like L (angles), C (channels) or tubes, are also used in construction when there are special requirements.

## Thin Walled Beams

A thin walled beam is a very useful type of beam (structure). The cross section of thin walled beams is made up from thin panels connected among themselves to create closed or open cross sections of a beam (structure). Typical closed sections include round, square, and rectangular tubes. Open sections include I-beams, T-beams, L-beams, and so on. Thin walled beams exist because their bending stiffness per unit cross sectional area is much higher than that for solid cross sections such a rod or bar. In this way, stiff beams can be achieved with minimum weight. Thin walled beams are particularly useful when the material is a composite laminates.

## Plate (Structure)

Folded structures are spatial structures formed by the elements in the plane, different in form and materialization. Folded structures differ in: geometric form, the form of a base over which they are performed, the manner of performance, methods of forming stiffness, function and position in the building, and the material they are made of.

By using folded structures different spatial forms can be made. The straight elements forming a folded construction can be of various shapes: rectangular, trapezoidal or triangular. By combining these elements we get different forms resulting in a variety of shapes and remarkable architectural expression.

Based on geometric shape folded structures can be divided into:

- Folded plate surfaces,

- Folded plate frames,

- Spatial folded plate structures.

Folded structures in the plane are the structures in which all the highest points of the elements and all the elements of the lowest points of the folded structure belong to two parallel planes.

Frame folded structures represent constructional set in which the elements of each segment of the folds mutually occupy a frame spatial form. This type of folded structure is spatial organization of two or more folds in the plane.

Spatial folded structures are the type of a structure in which a spatial constructive set is formed by combining mutually the elements of a folded structure.

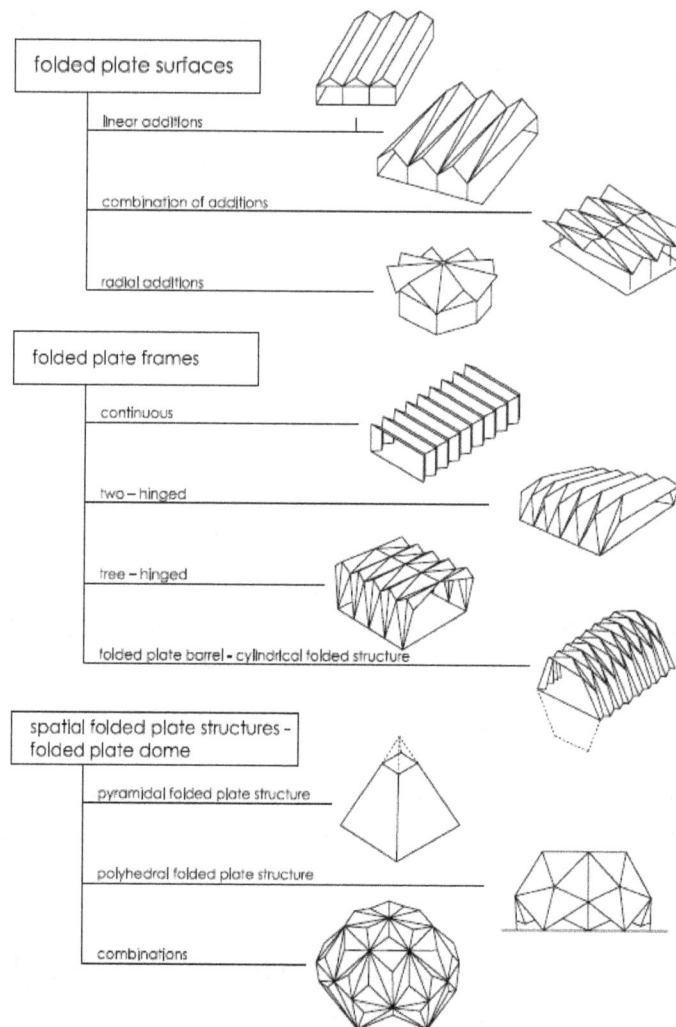

folded plate surfaces

linear additions

combination of additions

radial additions

folded plate frames

continuous

two – hinged

tree – hinged

folded plate barrel - cylindrical folded structure

spatial folded plate structures - folded plate dome

pyramidal folded plate structure

polyhedral folded plate structure

combinations

Figure: Forms of folded structures

The shape of folded structures affects the transmission of load and direction of relying of folded structures. Based on these parameters we can do the division in:

- linear folded plate structure,

- radial folded plate structure,

- spatial folded plate structure.

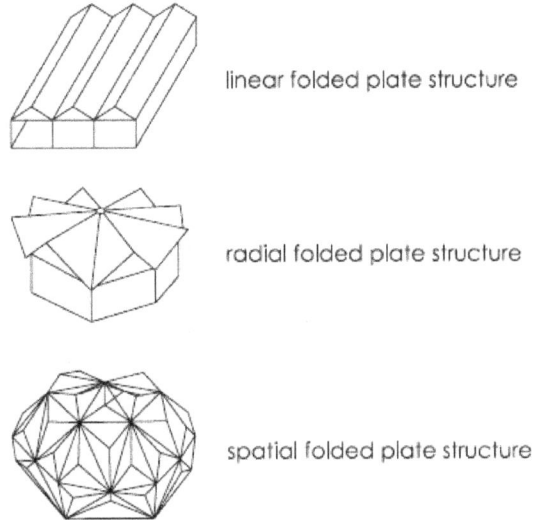

linear folded plate structure

radial folded plate structure

spatial folded plate structure

Fig.: Division of folded structures according to the direction of relying

Combined folded constructions are carried out over the complex geometric basis, formed by the combination of simple geometric figures, rectangles and semicircles on one side or both sides. This type of folded structure can be derived in the plane or as a frame (cylindrical) structure, and represents a combination of folded structure above the rectangular base and ½ of the radial construction.

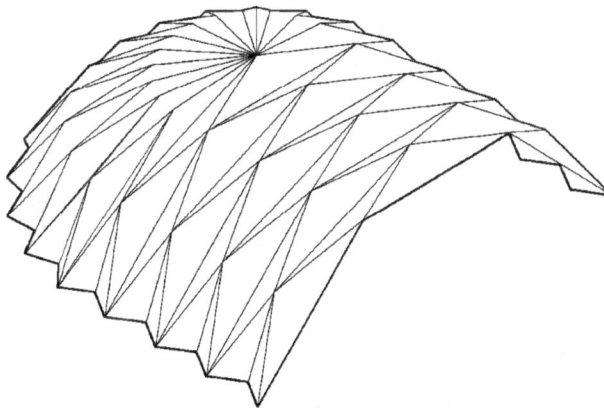

Fig.: Example of a combined folded structure formed by a cylindrical folded structure and ½ of a dome structure

Folded structure performed over an irregular basis is a complex spatial form. What characterizes this type of folded construction is the change of a range of the folded structure accompanying the base width. During the formation of such folded structures each triangular element of a fold has a different shape (size). This type of folded structure is formed by adjusting cylindrical folded structure to the irregular shape of the base.

By the method of execution folded structures can be divided into structures that are performed on site and prefabricated structures, which are produced in factories and assembled on site. If it is a large span structure, the elements are not fully performed in the factory, but it is imposed as a solution the performance of elements in the form of smaller segments for easier transport and installation on site.

## Diaphragm

Horizontal distribution of lateral forces to bents and shear walls is achieved by the floor and roof systems acting as diaphragms.

To qualify as a diaphragm, a floor or roof system must be able to transmit the lateral forces to bents and shear walls without exceeding a horizontal deflection that would cause distress to any vertical element. The successful action of a diaphragm also requires that it be properly tied into the supporting framing. Designers should ensure this action by appropriate detailing at the juncture between horizontal and vertical structural elements of the building.

Diaphragms may be considered analogous to horizontal (or inclined, in the case of some roofs) plate girders. The roof or floor slab constitutes the web; the joists, beams, and girders function as stiffeners; and the bents and shear walls act as flanges.

Diaphragms may be constructed of structural materials, such as concrete, wood, or metal in various forms. Combinations of such materials are also possible. Where a diaphragm is made up of units, such as plywood, precast-concrete planks, or steel decking, its characteristics are, to a large degree, dependent on the attachments of one unit to another and to the supporting members. Such attachments must resist shearing stresses due to internal translational and rotational actions.

The stiffness of a horizontal diaphragm affects the distribution of the lateral forces to the bents and shear walls. For the purpose of analysis, diaphragms may be classified into three groups rigid, semi rigid or semi flexible, and flexible although no diaphragm is actually infinitely rigid or infinitely flexible.

A rigid diaphragm is assumed to distribute horizontal forces to the vertical resisting elements in proportion to the relative rigidities of these elements.

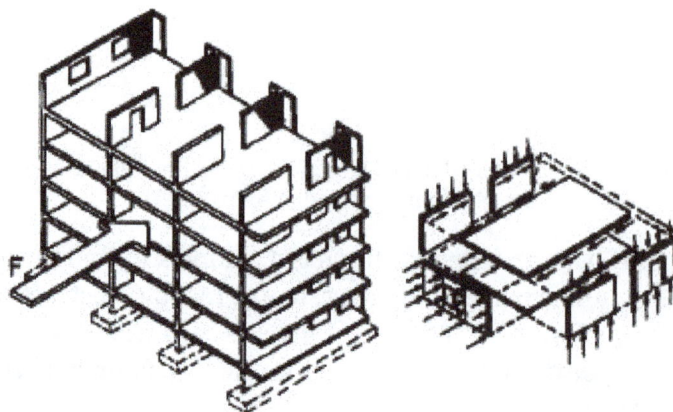

Figure: Floors of building distribute horizontal loads to shear wall.

Semi rigid or semi flexible diaphragms are diaphragms that deflect significantly under load, but have sufficient stiffness to distribute a portion of the load to the vertical elements in proportion to the rigidities of these elements. The action is analogous to a continuous beam of appreciable stiffness on yielding supports. Diaphragm reactions are dependent on the relative stiffnesses of diaphragm and vertical resisting elements.

A flexible diaphragm is analogous to a continuous beam or series of simple beams spanning between nondeflecting supports. Thus, a flexible diaphragm is considered to distribute the lateral forces to the vertical resisting elements in proportion to the exterior-wall tributary areas.

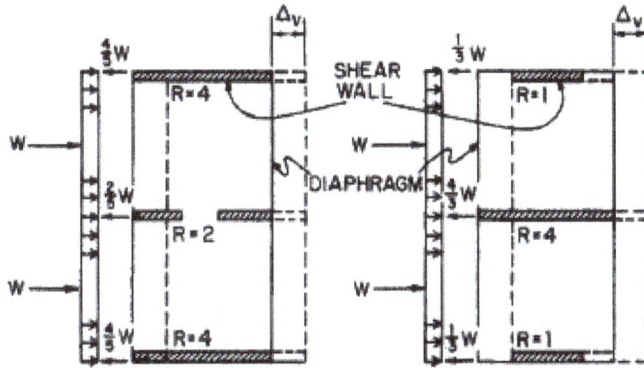

Figure: Horizontal section through shear walls connected by a rigid diaphragm. $R$ = relative rigidity and $\Delta_v$ = shear-wall deflection

STRUCTURAL THEORY

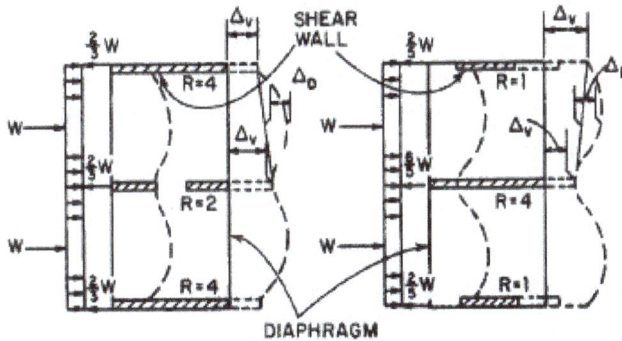

Figure: Horizontal section through shear walls connected by a semirigid diaphragm. $\Delta_v$ = diaphragm horizontal deflection

## Torque Distribution to Shear Walls

When the line of action of the resultant of lateral forces acting on a building does not pass through the center of rigidity of a vertical, lateral-force-resisting system, distribution of the rotational forces must be considered as well as distribution of the transnational forces. If rigid or semi rigid diaphragms are used, the designer may assume that torsional forces are distributed to the shear walls in proportion to their relative rigidities and their distances from the center of rigidity. A flexible diaphragm should not be considered capable of distributing torsional forces.

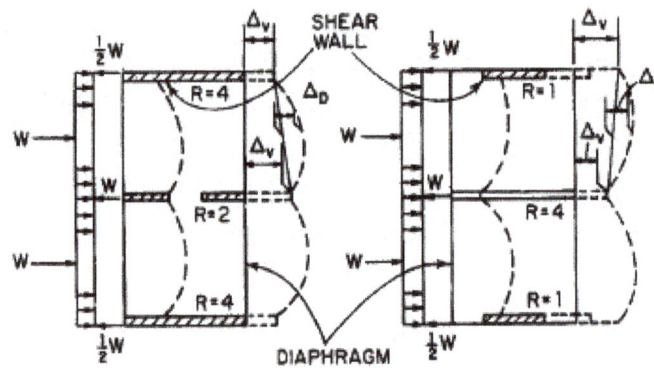

Figure: Horizontal section through shear walls connected by a flexible diaphragm.

## Deflections of Bents or Shear Walls

When parallel bents or shear walls are connected by rigid diaphragms and horizontal loads are distributed to the vertical resisting elements in proportion to their relative rigidities, the relative rigidity of the framing depends on the combined horizontal deflections due to shear and flexure. For the dimensions of lateralforce- resisting framing used in many high-rise buildings, however, deflections due to flexure greatly exceed those due to shear. In such cases, only flexural rigidity need be considered in determination of relative rigidity of the bents and shear walls.

Horizontal deflections can be determined by treating the bents and shear walls as cantilevers. Deflections of braced bents can be calculated by the dummy-unit load method or a matrix method. Deflections of rigid frames can be obtained by summing the drifts of the stories, as determined by moment distribution or a matrix method. And deflections of shear walls can be computed from formulas given, the dummy-unit-load method, or a matrix method.

For a shear wall with a solid, rectangular cross section, the flexural deflection at the top under uniform loading is given by the formula for a cantilever.

## Diaphragm Deflection Limitations

As indicated, horizontal deflection of diaphragms plays an important role in determining lateral-load distribution to bents and shear walls. Another design consideration is the necessity of limiting diaphragm deflection to prevent excessive stresses in walls perpendicular to shear walls. Equation was suggested by the structural engineers association of southern california for allowable story deflection, in, of masonry or concrete building walls.

This limit on deflection must be applied with engineering judgment. For example, continuity of wall at floor level is assumed, and in many cases is not present because of through-wall flashing. In this situation, the deflection may be based on the allowable compressive stress in the masonry, if a reduced cross section of wall is assumed. The effect of reinforcement, which may be present in a reinforced brick masonry wall or as a tie to the floor system in a nonreinforced or partly reinforced masonry wall, was not considered in development. Note also that the limit on wall deflection is actually a limit on differential deflection between two successive floor, or diaphragm, levels.

Maximum span-width or span-depth ratios for diaphragms are usually used to control horizontal diaphragm deflection indirectly. Normally, if the diaphragm is designed with the proper ratio, the diaphragm deflection will not be critical.

## Shear-Wall Rigidity

Where shear walls are connected by rigid diaphragms so that they must deflect equally under horizontal loads, the proportion of total horizontal load at any level carried by a shear wall parallel to the load depends on the relative rigidity, or stiffness, of the wall in the direction of the load. Rigidity of a shear wall is inversely proportional to its deflection under unit horizontal load. This deflection equals the sum of the shear and flexural deflections under the load.

Where a shear wall contains no openings, computations for deflection and rigidity are simple. Each of the shear walls has the same length and rigidity. So each takes half the total load. In figure, length of wall C is half that of wall D. By Eq. C therefore receives less than one-eighth the total load.

Walls with Openings. Where shear walls contain openings, such as doors and windows, computations for deflection and rigidity are more complex. But approximate methods may be used.

## Effects of Shear-Wall Arrangements

To increase the stiffness of shear walls and thus their resistance to bending, intersecting walls or flanges may be used. Often in the design of buildings, A-, T-, U-, L-, and I-shaped walls in plan develop as natural parts of the design.

Shear walls with these shapes have better flexural resistance than a single, straight wall.

In calculation of flexural stresses in masonry shear walls for symmetrical T or I sections, the effective flange width may not exceed one-sixth the total wall height above the level being analysed.

For unsymmetrical L or C sections, the width considered effective may not exceed one-sixteenth the total wall height above the level being analysed. In either case, the overhang for any section may not exceed six times the flange thickness.

The shear stress at the intersection of the walls should not exceed the permissible shear stress.

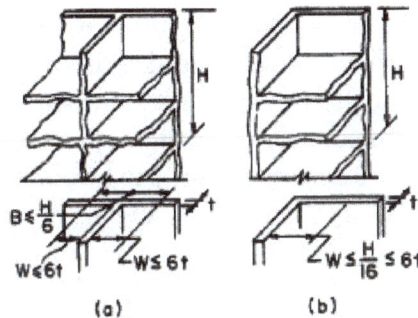

Figure: Effective flange width of shear walls may be less than the actual width:
(a) limits for flanges of I and T shapes: (b) limits for C and L shapes.

## Coupled Shear Walls

Another method for increasing the stiffness of a bearing-wall structure and reducing the possibility of tension developing in masonry shear walls under lateral loads is coupling of coplanar shear walls.

## Shell (Structure)

Shell elements take the form of planar, straight-edged triangles. When modeling curved shells, they provide a "faceted" approximation to the true geometry. Currently, the thickness of each element is assumed to be constant, although the thickness of adjoining elements may be different.

In all cases, six displacement degrees of freedom are used at each node. These include three translations parallel to the global axes and three rotations about these axes. For smooth shells it is possible to use only five degrees of freedom that is three translations and the two local rotations of the mid-surface normal. However, the use of six degrees of freedom in all cases has distinct advantages. Firstly, it allows non-smooth structures such as boxes, trays and stiffened shells to be analysed without applying any special constraint conditions to the junctions between intersecting surfaces. Secondly, beam elements (which in three-dimensional space must have six degrees of freedom) can be attached directly to the nodes of the underlying plate or shell mesh without applying constraints or transformations. Thirdly, in a large deflection (geometrically non-linear) analysis, the use of three rotational degrees of freedom allows an elegant and geometrically exact model for finite rigid body motion to be introduced. It is worth noting that many of the deficiencies of existing geometrical non-linear models of thin-walled structures can be traced to approximations made with respect to the rotational parameters. The parameters adopted here are "natural" in the sense that they are directly related to spectral analysis of the associated rotation tensor. To obtain linearized forms, use is made of the classical analysis of rigid body kinematics by Euler and Rodrigues.

The element formulations account for membrane, flexural and transverse shear deformations. The latter is based on the Reissner-Mindlin theory. In this theory the classical Kirchhoff assumptions for thin plates and shells are relaxed by lowering the continuity requirements from $C_1$ to $C_0$, including transverse shear. This allows both "thin" and "moderately thick" plates and shells to be modeled. In practice, the performance of such elements is known to deteriorate rapidly as the plate or shell becomes thin. This phenomenon is called shear locking and is caused by the inability of the element to approach the limiting condition of zero transverse shear strain at the appropriate quadratic rate. Shear locking is alleviated by the use of reduced integration, but contrary to early expectations, it is by no means eliminated. More recently, variants of the energy balancing techniques introduced by Fried and McNeal have produced excellent results over a wide range of structural aspect ratios (ratio of a characteristic dimension measured around the mid-surface of the shell structure to the average wall thickness). The form used for the present triangular elements is due to Tessler and Hughes, and allows accurate results to be obtained over the approximate aspect ratio range:

$$8 < r < 10^6$$

Apart from eliminating locking, the method reduces sensitivity of thin shells to element distortion, and improves the conditioning of the stiffness matrix and the quality of stress prediction.

Note: The analogous problem of membrane locking only occurs in thin curved elements and therefore need not be considered here.

Currently, only linear elastic materials can be modeled. These maybe either isotropic or orthotropic. In the latter case, the principal planes of the material are orthogonal, with one plane lying in the mid-surface of the shell and the other two planes intersecting this surface along two perpendicular lines referred to as the directions of orthotropic. The directions of orthotropic are determined by processing and are prescribed for each element from the Fill+Pack analysis. The material properties in each direction are specified by the user when preparing the analysis inputs file for the Warp or Stress analysis.

Geometric non-linearity is based on a convicted Lagrangian approach in which the displacement field is referred to a set of local convected coordinates that co-rotate with the cross-sections of the shell at each point. For consistency, the usual inextensibility condition is here applied to fibers that are collinear with the cross-sectional director at any point.

The co-rotational method derives from the polar decomposition theorem of continuum mechanics which asserts that any general spatial motion can always be decomposed into a pure stretch (deformation) followed by a rigid body motion. Adopting suitable finite rotation measures (see above) and explicitly discarding the rigid body component of the overall motion, a consistent method of evaluating the internal deformations and associated stresses within an element is achieved. In practice, this means that no limitations need to be placed of the magnitude of the rigid-body motion, and that the precision with which the internal deformations and stresses are determined will remain constant throughout the analysis.

The formulation is closely related to the finite deflection theories of Reissner, Simmonds and Danielson, Danielson and Hodges, and the finite element implementation of these theories by Bates and Simo and Vu-Quoc.

## Element LMT3

Element LMT3 is used in Midplane analysis technology. LMT3 is a three-noded triangular element with 18 degrees of freedom (six at each node). The element is constructed by superimposing the local membrane formulation due to Bergan and Nygard and Bergan and Felippa with the bending formulation due to Tessler and Hughes and transforming the combined equations to the global coordinate system. The resulting element can model bi-linear variations of membrane and transverse shear strains, but the flexural strains (curvatures) are constant.

The "free formulation" of Bergan and Nygard is based on assumed displacement fields, but goes beyond the strict potential-energy formulation in allowing "nonconforming" shape functions to be used. To ensure convergence, patch test requirements are enforced a-priori. The displacement shape functions are separated into basic and higher-order modes, the former being associated with rigid body and constant strain states and the latter with coordinate invariant in-plane bending modes.

This leads in turn to a basic and higher-order stiffness denoted by $[K_b]$ and $[K_h]$ respectively. The combined membrane stiffness then takes the form:

$$[K] = [K_b] + \beta[K_h]$$

where $\beta$ is a free parameter that acts as a scaling factor on the higher-order stiffness. In line with detailed test results, $\beta = 1.5$ has been adopted as the best choice.

The rotation about the local mid-surface normal (often referred to as the drilling freedom) is linked to the average in-plane rotation of the mid-surface by the penalty constraint technique. Thus,

$$\theta_z = \frac{1}{2}\left[\frac{\partial u_y}{\partial x} - \frac{\partial u_x}{\partial y}\right]$$

The drilling freedom is a fully integrated active component in the formulation. The strength of the link between $\theta_z$ and the in-plane gradients remains even in the case of exactly co-planar elements, so that singularity problems no longer occur. Great attention must be paid to the use of $\theta_z$ as a boundary condition component.

The basis of the bending formulation of the element is an explicit degree of freedom technique achieved via continuous transverse shear edge constraints. This leads to a constrained (coupled) total displacement field with bi-quadratic variation in the lateral displacement $u_z$ and bi-linear variations in the normal rotations $\theta_x$ and $\theta_y$. When combined with the transverse shear correction factor technique, the element exhibits a much improved flexural response (compared to the standard constant curvature formulation) across a wide range of aspect ratios.

## Element Definition

The geometry, node numbering and local/global degrees of freedom for the element are shown in The LMT3 Shell Element. Note that nodes i, j, k refer to entries in the nodal connectivity table given in the analysis output file. For example, if the connectivity for an element is 11, 101, 85 then i = 11, j = 101, k =85, and the local X-axis runs from node 11 to node 101.

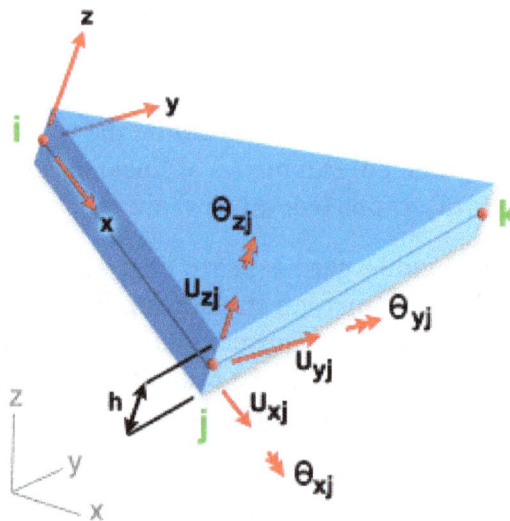

The LMT3 Shell Element

## Element Loading

Two types of element loading are available: pressure loading and initial strains due to orthotropic shrinkage.

Pressure loads are assumed to act on and be normal to the element mid-surface. The pressure is assumed to have a constant value over any given element, although pressure values used in adjacent elements may be different.

## Integration Rules

Integration over the mid-surface is carried out using three-point numerical quadrature. Since the material is linear elastic and the element is flat, both the strains and the stresses vary linearly through the shell walls. Consequently, explicit pre-integration in the thickness direction is used rather than the more expensive numerical integration.

The integration station locations and weights are shown in Integration station locations for LMT3 Element and Integration location weights for LMT3 Element.

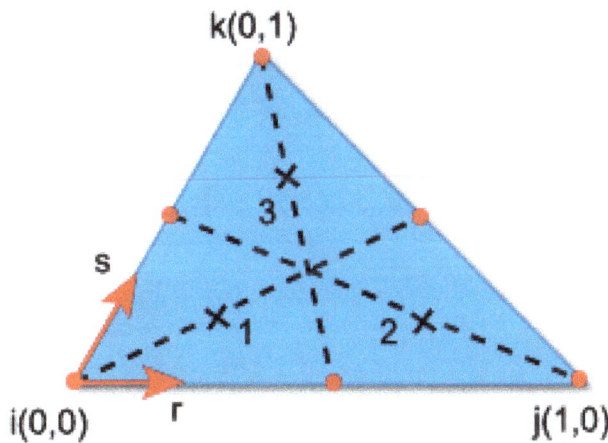

Integration station locations for LMT3 Element

## Integration location weights for LMT3 Element

| Element | Integration Station | r-coordinate | s-coordinate |
|---------|---------------------|--------------|--------------|
| LMT3    | 1                   | 1/6          | 1/6          |
|         | 2                   | 2/3          | 1/6          |
|         | 3                   | 1/6          | 2/3          |

## Results Output

Membrane forces $(N_x, N_y, N_{xy})$ bending moments $(M_x, M_y, M_{xy})$ and transverse shear forces $(Q_x, Q_y)$ are calculated at the integration stations and are illustrated in Results from Analysis for the LMT3 Element.

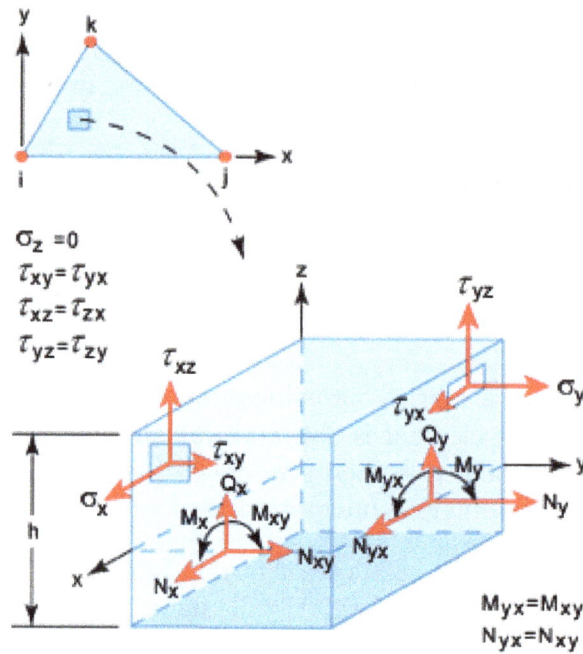

Assuming plane stress conditions, the stresses at any level $z$ are given by:

$$\sigma_x = \frac{N_x}{h} + \frac{12zM_x}{h^3}$$

$$\sigma_y = \frac{N_y}{h} + \frac{12zM_y}{h^3}$$

$$\tau_{xy} = \frac{N_{xy}}{h} + \frac{12zM_{xy}}{h^3}$$

$$\tau_{xz} = 1.5\frac{Q_x}{h}\left[1 - 4\left(\frac{z}{h}\right)^2\right]$$

$$\tau_{yz} = 1.5\frac{Q_y}{h}\left[1 - 4\left(\frac{z}{h}\right)^2\right]$$

Note that the forces and moments are those acting on a unit width of the plate or shell and so the membrane and shear forces have dimensions force/length, whereas the moments have dimensions (force).

## Column

A column is a vertical structural member intended to transfer a compressive load. For example, a column might transfer loads from a ceiling, floor or roof slab or from a beam, to a floor or foundations.

Columns are typically constructed from materials such as stone, brick, block, concrete, timber, steel, and so on, which have good compressive strength.

## Classical Stone Columns

In classical architecture, columns are often highly decorated, with standard designs including Ionic, Doric and Corinthian, and so on.

A colonnade is a row of columns spaced at regular intervals that can be used to support a horizontal entablature, an arcade or covered walkway, or as part of a porch or portico.

## Steel Columns

Steel columns have good compressive strength, but have a tendency to buckle or bend under extreme loading. This can be due to their:

- Length.

- Cross-sectional area.

- Method of fixing.

- Shape of the section.

The cross-sectional area and the section shape are incorporated into a geometric property of section, known as the radius of gyration. This refers to the distribution of an object's components around an axis. It can be calculated:

$$r = \sqrt{I/A}$$

Where, I = 2nd moment of area, A = cross-sectional area.

## Slenderness Ratio

The slenderness ratio is the effective length of a column in relation to the least radius of gyration of its cross-section. If this ratio is not sufficient then buckling can occur.

Column slenderness can be classified as:

- Long or slender: The length of the column is greater than the critical buckling length. Mechanical failure would typically occur due to buckling. The behavior of long columns is dominated by the modulus of elasticity, which measures a column's resistance to being deformed elastically (i.e. non-permanently) when a force is applied.

- Short: The length of the column is less than the critical buckling length. Mechanical failure would typically occur due to shearing.

- Intermediate: In between the long and short columns, and its behavior is dominated by the strength limit of the material.

Classification will depend on the column's geometry (i.e. its slenderness ratio) and its material properties (i.e. Young's modulus and yield strength).

## Shape

Columns can be classified according to their cross sectional shape. Common column shapes include:

- Rectangular.
- Square.
- Circular.
- Hexagonal
- Octagonal.

In profile, they can be tapered, non-tapered, or 'barrel' shaped, their surface can be plain, fluted, twisted, panelled, and so on.

Columns may be of a simple uniform design, or they may consist of a central 'shaft' sitting on a column base, and topped by a 'capital'.

## Reinforced Concrete Columns

Reinforced concrete columns have an embedded steel mesh (known as rebar) to provide reinforcement.

The design of reinforcement can be either spiral or tied.

- Spiral columns are cylindrical with a continuous helical bar wrapped around the column. This spiral provides support in the transverse direction.
- Tied columns have closed lateral ties spaced approximately uniformly across the column. The spacing of the ties is limited in that they must be close enough to prevent failure between them, and far enough apart that they do not interfere with the setting of the concrete.

## Other Types of Column

### Stone Column

Stone columns (or vibro stone columns) are formed by granular aggregate that is inserted into column shaped excavations and then compacted to improve the load bearing capacity of soil and fill material.

## Pilotis

Pilotis are supports that lift a building above the ground or a body of water. In timber form, they were traditionally used in the vernacular architecture of Asia and Scandinavia, or wherever indigenous peoples lived at a water's edge. They may also be used in hurricane or flood-prone areas, to raise the structure above storm surge levels.

The pioneer of modern pilotis was the architect Le Corbusier, who used them both functionally as ground-level supporting columns, and philosophically as a tool for freeing the rigidity of traditional plan layouts, enabling efficient, buildings as 'machines for living'.

## Piers

the term 'pier' can be used interchangeably for several different building elements. In general, it is an upright support for a structure or superstructure, but it can also refer to the sections of load-bearing structural walls between openings and different types of column.

## Posts

The word 'post' is interchangeable with the word column, although is is typically used in relation to smaller structural members that may in some cases be independent, rather than part of a wider structure.

## Tie

Ties are continuous tensioned reinforcements which are completely anchored and sufficiently lapped mechanically or using weld. There are various types of building ties used for various purposes.

## Types of Ties Used in Building Construction

Types of building ties include:

- Peripheral ties
- Internal ties
- Horizontal column and wall ties
- Vertical ties

## Peripheral Ties in Building Construction

Peripheral ties are commonly provided at roof and floor level and it placed at 1.2m away from the perimeter wall or edge of the building, as explained in figure.

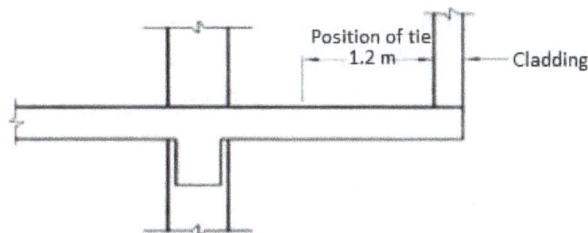

Figure: Location of Peripheral Ties

peripheral ties must withstand a tensile force equal to the lesser of 60KN or an amount computed according to the following equation.

$$F_t = 20 = 4n_o$$

Where:

n: is the number of storey of the structure

The area of steel bars required for the peripheral ties can be computed according to the following equation:

$$A_{st} = F_t \,/\, (0.87 f_y)$$

Where:

$A_{st}$: is the steel area required for peripheral ties

$F_t$: Tensile force that peripheral tie should resist

Peripheral ties need to be anchored and lapped adequately. The location of peripheral ties need for a building is illustrated in figure.

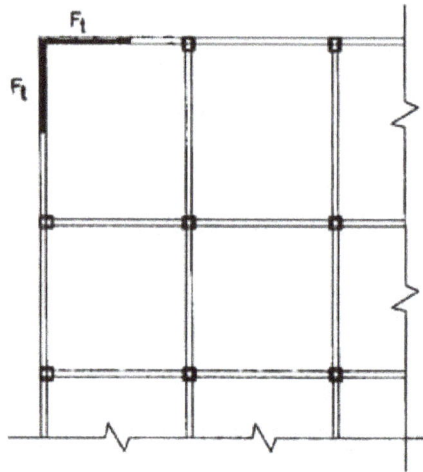

Figure: Location of peripheral ties are specified on typical floor plan

## Internal Ties in Building Construction

Internal ties are placed at roof level and floor level in two directions nearly perpendicular to each other.

Internal ties should be effectively continuous along their length and connected at both ends to the peripheral ties or anchored to the columns or perimeter walls when such ties are continuous to columns or perimeter walls. Figure illustrates the distribution of internal ties in typical floor plan.

The maximum distance between internal ties is equal to the 1.5 times the longest distance between centers of vertical loading elements in the direction of ties.

The internal ties must withstand a tensile force equal or greater than the force computed according to the following formula:

$$Tensile\ force = 0.0267 (g_k + q_k)\ lFt$$

Where:

$(g_k + q_k)$: is the sum of average characteristic dead load and live load exerted on the floor. These quantities are computed according to the specifications of Eurocode.

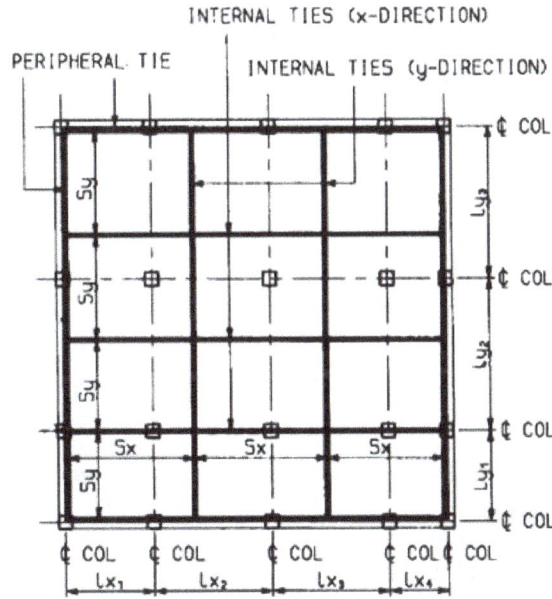

Figure: Distribution of Internal Ties in a Typical Floor of a Building

## Vertical Ties in Building Construction

It is specified to use vertical ties for buildings with minimum five storeys. Each wall and each column that support vertical loads need to be continuously tied from lowest level (foundation) to highest level (roof of the structure).

If the utilization of vertical ties is not possible, then the element must be designed in such a way that if such member is removed, then the surrounding elements should be designed to be able to bridge the gap and prevent failure due the removal of that element.

The force that vertical ties are subjected to is equal to the maximum design ultimate dead load and live load exerted on walls or columns from any one storey.

## Horizontal Ties to Column and Wall

These ties are used to connect external load bearing elements to the structure certain levels. Therefore, all external loads bearing members such as walls and columns are to be tied or anchored into structure at each roof or floor level horizontally.

The design tie force is equal to the greater of the two values computed according to the following expressions:

Design tie force $= 2\,F_t$ or $(\text{floor to ceiling height in m} / 2.5)\,F_t$

The smaller value is selected from equation

Design tie force = Three percent of the total ultimate vertical load in the wall or column at that level

Horizontal ties should be provided in two directions at approximately right angle for corner columns.

# Structural System

The Structural systems or structural frames can be defined as the assembly of inter-related or inter-dependent elements which forms a complex structure, and they are designed and built for resisting different loads.

The structural systems are the combinations of elements which serve a common purpose. The elements of structural systems can be related to the bones of the human body. If bones are weak and not properly aligned and integrated into the human body, then the human body would not be able to work or perform well. In the same way, if the structural systems are not proper then it would not be able to take loads.

The functions of Structural Systems are to resist loads acting on structures and provide a skeleton in the building which encloses and subdivides space to provide a protected environment.

A building can also be understood as a physical embodiment of a number of systems and subsystems integrated with each other forming building as a whole.

The structural system for a building particularly consists of a stable assembly of structural elements designed and constructed to support and transit applied loads safely to the ground where in its each member has a unique behavior under the applied loads.

The structural systems as the whole are divided in into different systems:

- Load Bearing System
- Framed System
- Shell System
- Strut and Tie

All the Structural Systems have different advantages and disadvantages. So depending upon the natural environmental condition and loading condition the proper structural system has to be chosen by the Structural Engineer.

The structural engineer should choose it based on the following properties:

- State of Equilibrium
- Adequate Strength
- Geometrical Stability
- Adequate Rigidity

## Substructure

The substructure is the portion of a structural system that is below the ground level. It includes foundations, columns, walls below ground level, basement slabs, bridge piers and abutments, and the base of retaining wall. The purpose of the substructure is to support superstructure and transfer the load coming from superstructure to ground below.

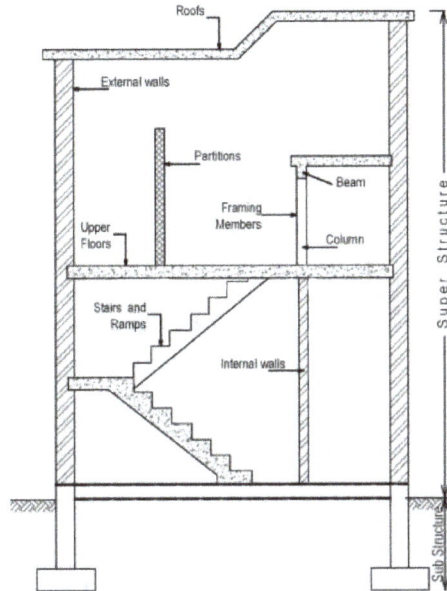

In the substructure, one of the most important elements is the foundation. The foundation system must be designed to transmit the loads from the superstructure structural system directly to the ground in such a manner that settlement of soil for the completed building will be within acceptable limits.

## Superstructure

The superstructure is the portion of a structural system above the ground surface and visible after completion of the project. It includes beams, slabs, columns, domes, shells and walls above ground, steel frames, bridge deck, and the wall of retaining wall etc.

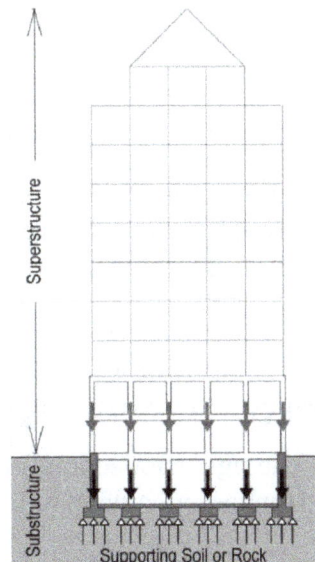

In most of the buildings, the superstructure consists of floor and roofs, horizontal members that support them and vertical members that support the other components. The superstructure is designed to transmit its loads to the foundation system in the manner anticipated in the design of the foundations.

The superstructure of the building is loaded with different loads and the different loading conditions. Hence instability of the structure may occur.

One way to provide lateral stability is to incorporate in the diagonal members called bracing in the structural system. By providing bracing, columns and beams work together to carry the lateral loads downward.

Another way is to rigidly connect beams to columns to prevent a change in the angle between the beams and columns, thus making them work together as a rigid frame to resist lateral movement.

Another method is to provide long walls known as shear walls in two perpendicular directions. Lateral forces on the building can be resolved into forces in each of these directions. The walls then act like vertical beams (cantilevers) in transmitting the forces to the foundations.

So from all the above discussion, it is seen that it is also important to protect the structural system against damages, including the fire. For fire protection, steel bracing can be encased in fire-resistant floors, roofs or walls. Similarly, columns can be encased in walls and beams may be encased in floors. Or fire-resistant materials, such as concrete, mineral fiber or plaster may be used to box in the structural members.

It is the choice of a right structural system which will render the optimum use of the building at an optimum cost. One has to choose the right structural system striking balance between its intended use and functions and of course the cost associated with it. Space, i.e. volume, height, column or obstruction free area, i.e. long span/short span etc. will decide the right structural system. And having chosen the right structural system, it is the structural/geometrical configuration of placing all structural elements, which will yield the most economical design, i.e. optimum cost of the building.

## Rigid Frame

Rigid frame offers a rotational stability that enhances how it carries vertical loads, increasing the longevity of the entire structure. Besides rigid frame, the same framing principles are used in the simple frame and partially restrained frame buildings.

Rigid frame is the beam to column connections are classified as rigid, hence the name. The frame is designed to transmit beam end moments and shear forces into the columns without bracing systems to resist lateral loads. The members can be straight or tapered.

The frame stability if only provided by the rigid connections and member stiffness. It looks similar to post and beam but is significantly stronger and able to hold vertical loads.

Local beam-column connection rotations are not considered in global frame analysis; the connections are designed to transmit the resulting beam end movements and shear forces into the columns.

The joints are not always fully fixed to either horizontal or vertical members; when the beam rotates from a vertical load, the columns rotate with it. This allows the joint to rotate as a unit and members maintain the same angular relationships during the rotation.

Rigid frame buildings are highly adaptable and flexible in design. Doors and windows can be placed anywhere, and HVAC units can be placed on the roof or the side. The exterior can easily be dressed to look like any envelope type including stone, brick, or wood.

## Differences Between Post and Beam and Rigid Frame

- Post and Beam: When a vertical load weighs on a common post and beam structure, it is carried by a horizontal member and then shifted by bending to columns or vertical members. The beam is simply supported by its columns, sitting on top of them so that the ends of the beams can rotate on top of the columns with no restraint. The result is that the horizontal members of the structure then only carries axial forces.

- Rigid Frame: When a rigid frame structure is subjected to a vertical load, it is also picked up by the beams and eventually transferred through the columns to the ground. However, the joints are strongly connected, preventing any free rotation from occurring at the beams ends. This slight difference changes everything about the behavior of the beams, which is now the same as a fix-ended beam.

Rigid frame construction provides many benefits, such as decreased deflections, decreased internal bending moments, and increased rigidity. However, the columns are experiencing some degree of internal bending themselves as the beams stay rigid.

A rigid frame structure can be designed smaller than other post and beam systems. This is because internal bending moments are reduced by the rigidity. On the other hand, the columns or vertical members should be designed to be a bit larger, since they are carrying both axial loads and internal bending moments.

## Moment Resisting Connections

The so-called rigid connections are typically full depth end-plate connections and extended end-plate connections. The most common of these is the bolted end-plate beam-to-column connection. The selection depends on the budget.

Welded connections can be used in place of the bolted end-plate, especially in seismically active regions. Welded connections can provide full moment continuity but tend to be on the expensive side. If used, the welded connections should be prefabricated rather than welded on-site.

Site-bolting takes less time and minimizes labor costs.

## Building Applications

Rigid frame is found in a wide variety of building styles and uses:

- Warehouses

- Retail stores

- Churches

- Plants

- Agricultural buildings

- Equipment shelters

- Multi-story buildings of any height

Rigid frame buildings are typically used when there are special requirements such as medical centers, research facilities, white rooms, and structures housing equipment sensitive to vibrations and deflection.

## Benefits of Rigid frame Construction

Beyond the clear span capabilities, which provide open spaces with no center columns or bracing systems, rigid span steel building retain all the benefits of any metal building.

- Cost-effective

- Energy-efficient

- Floors are not sensitive to vibration

- Connections perform better in load reversal situations and earthquakes

There are a few disadvantages in that the connections are more complex and can complicate the erection process. The initial cost of the structure is greater as well, but the investment is easily returned by the long life of the building.

## Braced Frame

A braced frame is a structural system commonly used in structures subject to lateral loads such as wind and seismic pressure. The members in a braced frame are generally made of structural steel, which can work effectively both in tension and compression.

The beams and columns that form the frame carry vertical loads, and the bracing system carries the lateral loads. The positioning of braces, however, can be problematic as they can interfere with the design of the façade and the position of openings. Buildings adopting high-tech or post-modernist styles have responded to this by expressing bracing as an internal or external design feature.

## Bracing Systems

The resistance to horizontal forces is provided by two bracing systems:

## Vertical Bracing

Bracing between column lines (in vertical planes) provides load paths for the transference of horizontal forces to ground level. Framed buildings required at least three planes of vertical bracing to brace both directions in plan and to resist torsion about a vertical axis.

## Horizontal Bracing

The bracing at each floor (in horizontal planes) provides load paths for the transference of horizontal forces to the planes of vertical bracing. Horizontal bracing is needed at each floor level, however, the floor system itself may provide sufficient resistance. Roofs may require bracing.

## Types of Bracing

## Single Diagonals

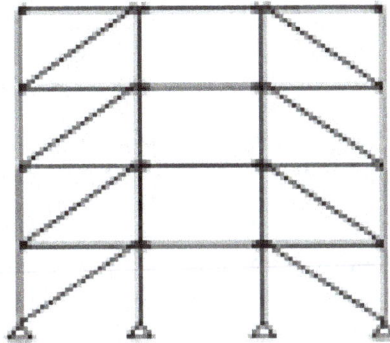

Trussing, or triangulation, is formed by inserting diagonal structural members into rectangular areas of a structural frame, helping to stabilize the frame. If a single brace is used, it must be sufficiently resistant to tension and compression.

## Cross-bracing

Cross-bracing (or X-bracing) uses two diagonal members crossing each other. These only need to be resistant to tension, one brace acting to resist sideways forces at a time depending on the direction of loading. As a result, steel cables can also be used for cross-bracing.

However, this provides the least available space within the façade for openings and results in the greatest bending in floor beams.

## K-bracing

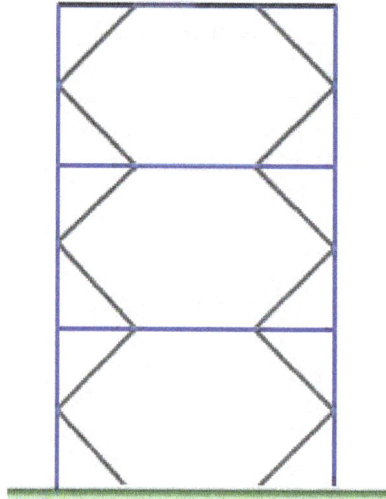

Braces connect to the columns at mid-height. This frame has more flexibility for the provision of openings and results in the least bending in floor beams. K-bracing is generally discouraged in seismic regions because of the potential for column failure if the compression brace buckles.

## V-bracing

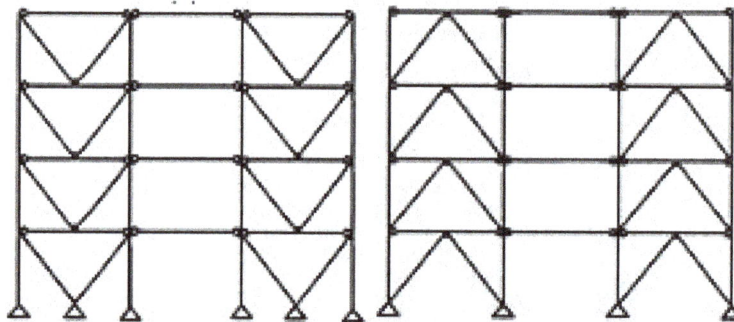

This involves two diagonal members extending from the top two corners of a horizontal member and meeting at a center point at the lower horizontal member, in the shape of a V. Inverted V-bracing (also known as chevron bracing) involves the two members meeting at a center point on the upper horizontal member.

Both mean that the buckling capacity of the compression brace is likely to be significantly less than the tension yield capacity of the tension brace. This can mean that when the braces reach their resistance capacity, the load must instead be resisted in the bending of the horizontal member.

## Eccentric Bracing

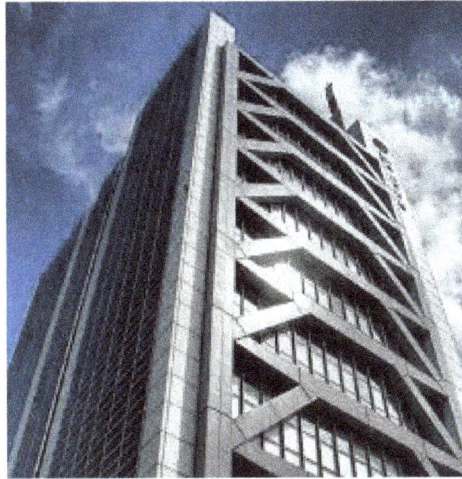

This is commonly used in seismic regions. It is similar to V-bracing but instead of the bracing members meeting at a center point there is space between them at the top connection. Bracing members connect to separate points on the horizontal beams. This is so that the 'link' between the bracing members absorbs energy from seismic activity through plastic deformation. Eccentric single diagonals can also be used to brace a frame.

## Space Frame

A space frame or space structure is a truss-like, lightweight rigid structure constructed from interlocking struts in a geometric pattern. Space frames can be used to span large areas with few interior supports. Like the truss, a space frame is strong because of the inherent rigidity of the triangle; flexing loads (bending moments) are transmitted as tension and compression loads along the length of each strut.

### Space Frame Applications

The integrated terminal buildings are designed with a number of structural innovations. The key feature of airport structure is their long span roof covering with massive column spacing.

- Commercial and industrial structures
- Auditoriums
- Sky lights
- Canopies
- Toll booths
- Exhibition halls
- Sports stadiums

## Edge of Space Frames

- The quality of roofs of this system is extremely good and the chances of deflection is almost nil;

- The installation of the structure can be done with the help of a mobile scaffold in places where a crane cannot be used. Since light weight tube elements & spherical balls forming the space frame system, it can be lifted by human effort;

- The space frames can cross large spans in both directions unlike the conventional ones. This gives great versatility and freedom to the designer in the selection of the structural system;

- The 3d system used in the system gratification ensures the correct utility rate of materials which ensures huge saving in steel usage;

- Since the system is prefabricated under factory supervision one can expect the rightful quality control;

- Steel or concrete substructures can easily be integrated with the Space frames;

- Minimal erection time as every component is fabricated;

- Storage and transportation of the system components are quite economical;

- The actual weight of steel membranes for large spans are less;

- The multi-directional structural behavior grants unusual design;

- The system is very convenient to fix any kind of cladding on the top by bolting down into the threaded hole provided at the top of each nodal joint;

- The system offers flexibility to fix the light fixtures and ceiling systems;

- By this method large span and big areas can be installed in shortened time period;

- Since the self-weight of space frame is small, greater economy is gained in the substructures as an added advantage;

- It is possible to integrate space frames in the architectural designs of various building types;

- When required the space structure system can be easily demounted and rebuilt them at another location;

- Space frame structures can be cladded with sandwich panels, acrylic, polycarbonate or glass glazing and will present a very modern and aesthetic view from both indoors and outdoors;

- In regard to attachments, it can be cladded with galvanized steel, aluminum, skylights, sandwich panels, fabric structures etc.

## Advantages of the 3D Space Frame Structures

Structural efficiency, rigidity, reduced deflections, integration of services, resilience, ease & speed of manufacturing and installation are some of the biggest advantage of 3D Space Frame Structures.

- Due to the 3D nature of space frame system long and clear spans up to 180 m can be easily achieved in lightweight steel construction, whereas this is extremely difficult in conventional structures, if done with difficulty they will be heavier by two or three fold compared to space frames.

- Since space frames are factory produced as light weight components maximum 3.5 m length, they can be Hot Dip Galvanized in Molten Zinc bath and over the galvanized surface they are further protected by either by powder coat or liquid spray paint to protect the entire structure from corrosion especially when serviced in extremely corrosive environment ( i.e. Corrosion Class 4 (Near Sea, Chemical Industries etc.), this coating system can warrant for a period of 20 years for first maintenance of the coating work on structure.

- Conventional trusses, steel sections used are longer compared to Space frame components so Hot Dip Galvanizing and powder coating on these items are often not practical hence their corrosion resistance is weak demands regular maintenance and may lead to premature failures.

- The open nature of the structure between the two plane grids allows easy installation of mechanical and electrical services and air-conditioning ducts, sprinkler system, firefighting services etc. within the structural depth of the modular construction.

- Their fixing is simplified as there is a regular grid of supports available thus reducing or even eliminating the need for secondary steelwork to support them.

## Protection Against Corrosion and Fire

Space structure system appears as structurally sealed circular hollow sections concentrically connected to solid spherical steel nodes stands out as a superior aesthetic and architectural end product. The exposed surfaces of space frame members are powder coated to decorative color finish in a factory controlled environment. The most modern and automated zinc phosphate treatment/powder coating technology provides weather and abrasion resistant thermoset polyester coatings in various eye catching colors selected from German color Standard RAL Or Internationally accepted NCS.

When specified to be fire protected, the exterior exposed surface of the components is spray coated with UL certified intumescent paint for minimum one hour fire rating. In such cases, the space structure is designed to select the optimum size and thickness of each element in the structure for the required HP/A value that is Heated Perimeter to Cross sectional Area to achieve a minimum fire rating of 60 minutes. If there is no specific fire rating required, then the structure designed as standard space frame as per project specified loadings as well as design codes and a UL certified intumescent paint is coated as recommended by the coating manufacturer.

## Design and Installation

In-house developed structures by the expert companies are so designed nowadays that it can perform specified tasks of structural analysis to international codes AISC, ANSI, ASCE, etc. as well it can generate manufacturing data and installation drawings to ensure speed and accuracy throughout construction.

The method chosen for the installation of space frame depends on its behavior of load transmission and construction details, so that it will meet the overall requirements of quality, safety, speed of construction and economy. The options available are three.

- Scaffold Method: Individuals Elements are assembled in Place at actual Elevations, members and joints or prefabricated subassembly elements are assembled directly on their final position.

- Block Assembly Method: The Space frame is divided on its plan into individual strips or blocks these units are fabricated on the ground level, then hoisted up into final position and assembled on the temporary supports.

- Lift-up Method: The whole space frame is assembled at ground level so that most of the work can be done before hoisting.

## Advantage of Space Frame Structure over Conventional and Pre-Engineering Steel Roof Structure

- Space Frame Structures can be fixed and dismantled as per need and desire;

- The structure is beautiful in look and in case of structure it is very safe for shorter to longer period;

- This structure is every architectures delight as this system, enables work of any kind of geometrical shape and profile;

- In this structure the distribution if lighting is very convenient as there is a presence of nodes and balls over the full complete area. In this system, AC ducts, Fire Sprinkler system can pass through the without requiring additional space and false ceilings;

- Load transfer in this structure is completely apt due to concentric and zero eccentric connections in axials;

- All the space frames pipes( hollow space) and components are perfectly protected against atmospheric exposure and attacks in this type of structure;

- For better corrosion resistance in the system, Galvanized product can be obtained with powder coating;

- This structure is anti-quake as the dead loads (Mass Lumps) are distributed over many joints/nodes which lead to no or minimal loss of the structure during tremors and earthquakes;

- Structural integrity and global ductility is superior in this system;

- This structure is capable of taking heavier loads from roof of steel roofed buildings;

- The space frame structure provides wide working floor area without any obstruction through intermediate supporting columns;

- This system is highly suited for shopping malls, Auditoriums, Function Halls, Stadiums, factory buildings, pedestrian bridges, Railway stations, Airport terminal buildings etc.;

- This system when adopted for small structures like Pargolas , Entrance gate arches, the structure gives elegant and rich look with excellent aesthetic appearance;

- The space frame structure is considered as hinged and hence supporting media (Concrete or steel columns and foundations) are economical as there is no moment transfer from SF to Supporting media.

## Shear Wall

Shear wall is a structural member in a reinforced concrete framed structure to resist lateral forces such as wind forces. Shear walls are generally used in high-rise buildings subject to lateral wind and seismic forces.

In reinforced concrete framed structures the effects of wind forces increase in significance as the structure increases in height. Codes of practice impose limits on horizontal movement or sway.

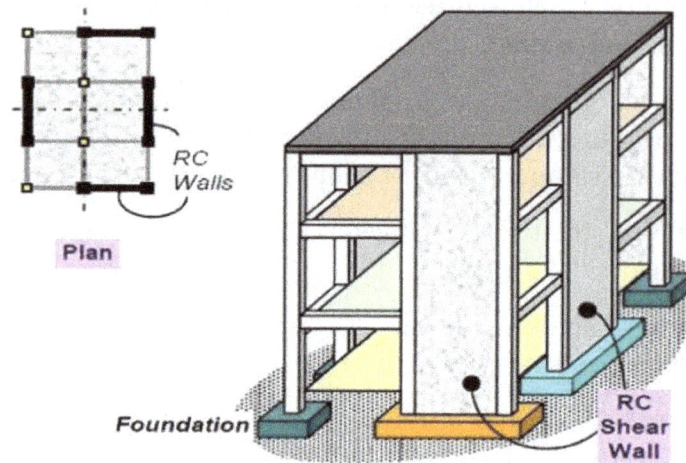

Limits must be imposed on lateral deflection to prevent:

- Limitations on the use of building,
- Adverse effects on the behavior of non-load bearing elements,
- Degradation in the appearance of the building,
- Discomfort for the occupants.

Generally, the relative lateral deflection in any one storey should not exceed the storey height divided by 500.

The figure below shows the deflected profiles for a shear wall and a rigid frame.

One way to limit the sway of buildings and provide stability is to increase the section sizes of the members to create a rigid, moment-resisting frame.

However, this method increases storey heights, thus increasing the building cost. It is rarely used for more than 7 or 8 storeys.

Another way is to provide stiff, shear resisting walls linked to a flexible frame. These can be external walls or internal walls around lift shafts and stairwells (a core) or sometimes both are provided.

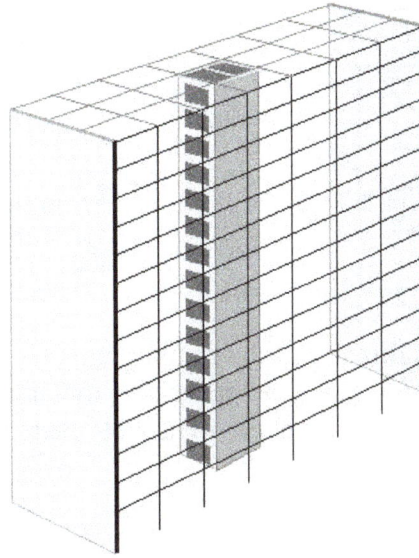

## Structural Forms or Types of Shear Walls

Monolithic shear walls are classified as short, squat or cantilever according to their height to depth ratio.

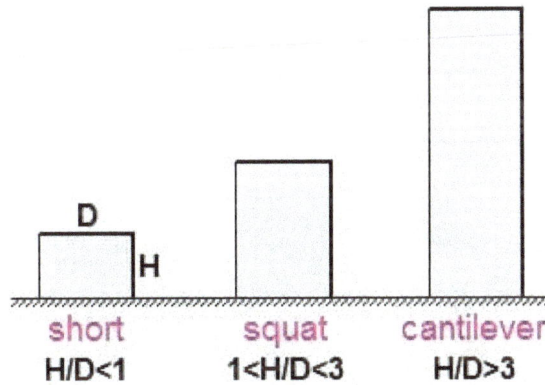

| short | squat | cantilever |
|-------|-------|------------|
| H/D<1 | 1<H/D<3 | H/D>3 |

Generally shear walls are either plane or flanged in section, while core walls consist of channel sections.

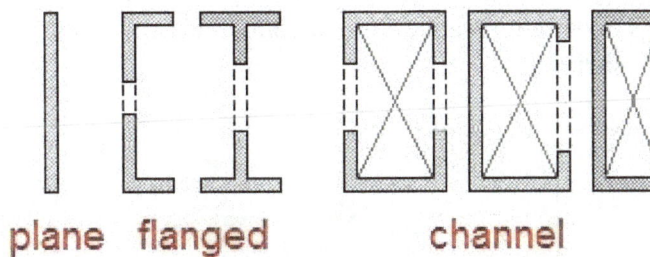

plane   flanged          channel

In many cases, the wall is pierced by openings. These are called coupled shear walls because they behave as individual continuous wall sections coupled by the connecting beams or slabs.

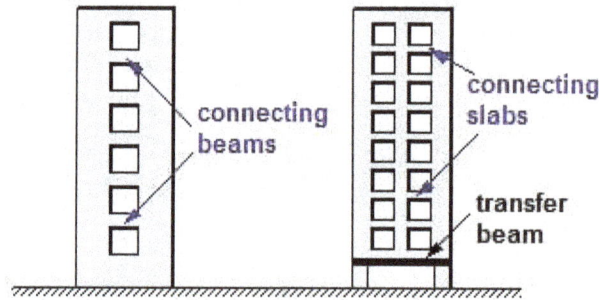

Normally the walls are connected directly to the foundations. However, in a few cases where the lateral loads are relatively small and there no appreciable dynamic effects, then they can be supported on columns connected by a transfer beam to provide clear space.

## Location of Shear Walls in a Building

The shape and plan position of the shear wall influences the behavior of the structure considerably. Structurally, the best position for the shear walls is in the center of each half of the building. This is rarely practical, however, since it dictates the utilization of the space, so they are positioned at the ends.

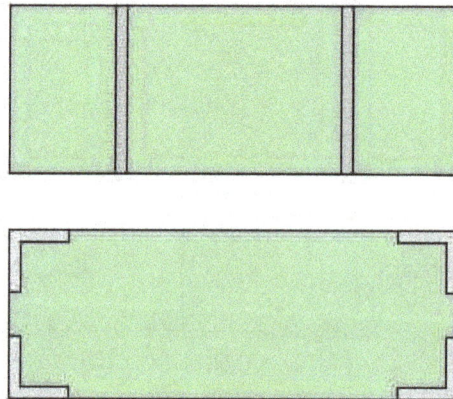

This shape and position of the walls give good flexural stiffness in the short direction, but relies on the stiffness of the frame in the other direction.

This arrangement provides good flexural stiffness in both directions, but may cause problems from restraint or shrinkage. As does this arrangement with a single core, but which does not have the problem from restraint of shrinkage.

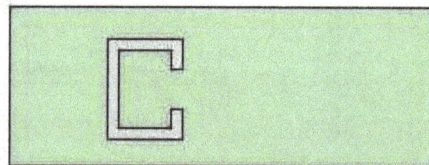

However, this arrangement lacks the good torsional stiffness of the previous arrangements due to the eccentricity of the core.

If the core remains in this position then it must be designed explicitly for the torsion. It is far preferable to adopt a symmetrical arrangement to avoid this.

## Tube

The tubular system is to arrange the structural elements in such a way that the system can resist the imposed loads on the structure efficiently especially the lateral loads. This system comprises of various elements i.e. slabs, beams, girders, columns. The walls and cores are engaged to resist the lateral loads, in the tubular system the horizontal loads are resisted by column and spandrel beams at the perimeter level of the tubes.

Types of tubular systems:

    (a) Framed tube structure

    (b) Braced tube structure

    (c) Tube in tube structure

    (d) Bundled tube structures

(a) Framed tube structures: Frames comprises of closely spaced columns, 2 to 4 m between centers, with deep girders joining them. The ideology is to develop a tube like structure which acts like a continuous perforated chimney or stack. The lateral resistance of this structure is provided by stiff moment resisting frames which form a tube throughout the periphery of the building. The gravity loads are distributed between the tube and the interior columns. This structural form provides an efficient structure appropriate for buildings with 40 to 100 storeys. When the horizontal loads act, the boundary frames arranged in the load direction acts as the webs of massive tube cantilever while those normal to the direction of the loading act like flanges. Although framed tube is the most structurally efficient system, flange frames will suffer from shear lag. This causes the mid face flange columns subjected to less stress than that of corner columns and so they won't contribute to their full potential strength. Example: Aon Centre and W.T.C Towers.

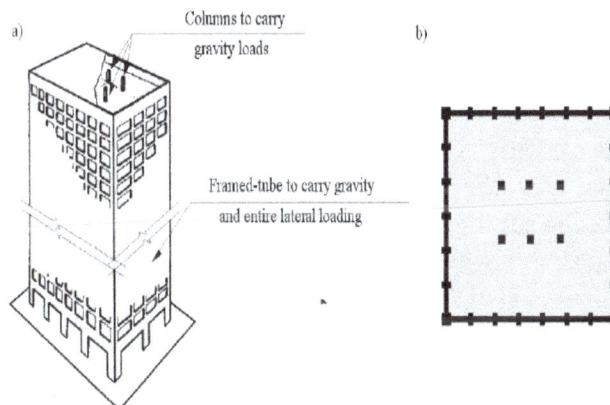

Figure: a) 3D view of Frame tube structure and b) sectional plan

The framed tube structure is shown in figure can be considered to be composed of: (1) Two web panels parallel to the direction of the lateral load; (2) two flange panels normal to the direction of the lateral load. These structural components are interconnected to each other along the panel joints and connected to the floor slabs at each floor level. The high in-plane stiffness of the floor slabs will restrict any tendency for the panels to deform out-of-plane and it may, therefore, be assumed that the out-of-plane actions are insignificant compared to the primary in-plane actions. If the sizes and spacing of the frame members are assumed uniform, as is usually the case in practice, then each framework panel may be replaced by an equivalent uniform orthotropic membrane.

(b) Braced tube structures: The tubular structure is further improved and can be done by cross bracing the frame with X-bracings throughout the entire building. As the braced tube diagonals are connected to the column at each and every intersection, they virtually erase the shear lag effects in flange and web frames together. As a result, the structure behaves like a braced frame under lateral loads by reducing bending in the frame members. Example: John Hancock Building.

Hence spacing between the columns shall be increased and the depth of girders can be made less, which facilitates large size windows unlike in conventionally framed tube structures. In braced tube structures, the braces are provided to share the axial load from more highly stressed columns to less highly stressed columns and this phenomenon helps to lower the difference between load stresses in columns. Example: Chicago's John Hancock building, The Citigroup Center, Bank of China Tower.

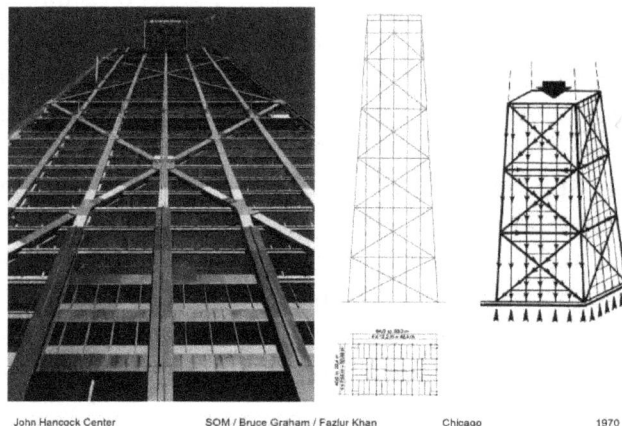

John Hancock Center          SOM / Bruce Graham / Fazlur Khan          Chicago          1970

Figure: 3D view of Braced tube structure and sectional plan

(c) Tube- in -tube structures: This is another type of framed tube consisting of an outer-framed tube along with an internal elevator and service core. The inner tube consists of braced frames. The outer and the inner tubes act together to resist both gravity and lateral loads in steel framed buildings. However, outer tube always plays an important role because of its greater structural depth. This type of structures is also referred as hull and core structures.

Tube-in-tube building generally consists of an inner tube to aid vertical transportation demand and an outer tube which comprises of dense columns and deep beams. It is the most commonly used structural system for high-rise building with more than 50 storeys. In order to facilitate the computational efficiency in the preliminary design, a numerous approximate analysis approaches were proposed to substitute the Finite Element Method which is elaborate but too exhaustive to calculate. Most of the approximate analysis for horizontal vibration analysis considers that the tube-in-tube structure is a double cantilever beam system with acceptable deformation between the two tubes. On the basis of continuum parameter technique, accurate solution of the double beam system is obtained, especially when structural parameters are assumed to be the same along the height of structure.

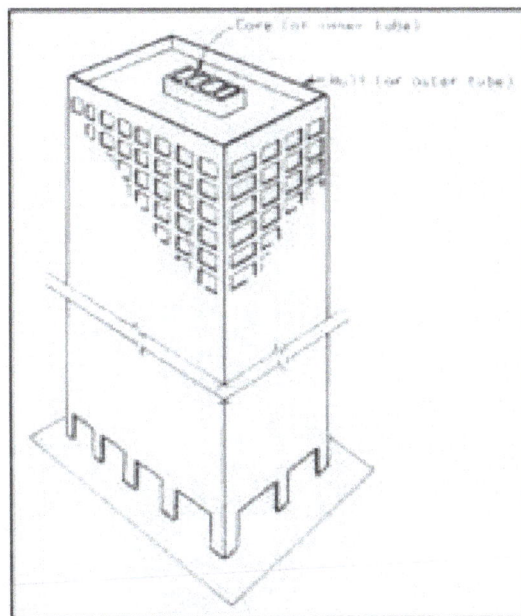

Tube-in-tube

Figure: 3D view of Tube in Tube structure

(d) Bundled tube: The bundled tube system can be characterized as an assemblage of individual tubes which results in multiple cell tube. This System allows for great heights and large floor area.

In this system, internal webs if introduced will greatly reduce the shear lag in the flange beams. Hence their columns are more uniformly stressed than in the single tube structures and they contribute more to the lateral stiffness. Example: Sears Tower.

## Advantages of Tubular System

It offers some clear advantage from materials standpoint. Designed well, tubular forms have been known to utilize the same amount of material as would have been employed for a structure that is half as large or framed conventionally.

- Allows greater flexibility in planning of interior space since all the columns and lateral system is concentrated on the perimeter of structure. This allows a column-free space in the interior.

- Regularity in the column schedule allows off-site fabrication and welding where speed can be achieved while still confronting to quality.

- Wind resisting system since located on the perimeter of the building meant that maximum advantage is taken of the total width of the building to resist overturning moment.

- Identical framing for all floors because floor members are not subjected to varying internal forces due to lateral loads.

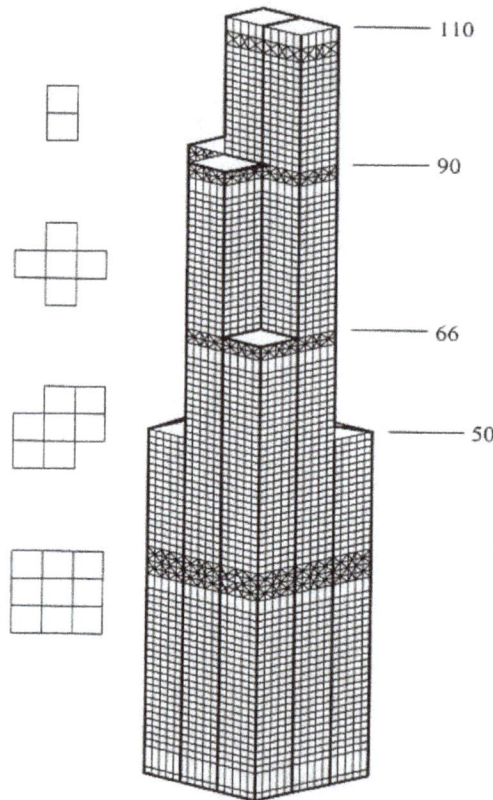

Figure: 3D view of Bundled Tube structure

# Eyebar

An eyebar is a structural member having a long body and an enlarged head at each end, each head has a hole through which a pin is inserted to connect to another member. Eyebars are tension only members. Eyebars are predominantly found on older pin connected truss bridges. Eyebars are used as different members of a pin connected truss bridge; the most common member used is diagonals, diagonal counter, and lower cords. They can be found used as vertical hangers and suspension chains.

Eyebars are a fracture critical member of the truss, so they should receive special attention during the inspection process of the structure. Many pin-connected trusses using an eyebar as a member are non-redundant, having only one or two eyebars per member. A redundant eyebar member will consist of three or more eyebars.

The detail of the eyebar may be designed to different sizes and lengths, depending on the location in the truss. The different size eyebar will experience different level of tension. This tension can be measured by counting the oscillation vibration of the bar under a dead load.

## The Major Advantage of Eyebars

The major advantages of using pin connection detail are the design simplicity and the ability for free end rotation. The design simplicity afforded by pin connections reduces the amount and complexity of design calculation.

By allowing for free end rotation, pin connections reduce the level of stress in the member. This free end rotation allows for the use of less steel in the connection detail. This was important to early steel bridge designers when steel costs were at a premium.

## The Major Disadvantage of Eyebars

The major disadvantages of pin connection details are the result of vibration, pin wear, eye wear, unequal eyebar tension, unseen corrosion, and poor inspect ability. Vibrations increase with pin connection because they allow more movement than rigid types of connections. When there is an increase in pin or eyebar wear, the eyebar will decrease in tension, causing the eyebar to have larger longitudinal side to side vibration under a live load. As a result of this side to side vibration, the moving parts are subject to wear.

## Tensile Structure

Tensile structure is the term usually used to refer to the construction of roofs using a membrane held in place on steel cables. Their main characteristics are the way in which they work under stress tensile, their ease of pre-fabrication, their ability to cover large spans, and their malleability. This structural system calls for a small amount of material thanks to the use of thin canvases, which when stretched using steel cables, create surfaces capable of overcoming the forces imposed upon them.

There are three different main classifications in the field of tensile construction systems: membrane tensioned structures, mesh tensioned, and pneumatic structures. The first relates to structures in which a membrane is held by cables, allowing the distribution of the tensile stresses through its own form. The second case corresponds to structures in which a mesh of cables carries the intrinsic forces, transmitting them to separate elements, for example, sheets of glass or wood. In the third case, a protective membrane is supported by means of air pressure.

Structurally, the system is formalized by combining three elements: membranes, rigid structures such as pole and masts, and cables.

The membranes of PVC-coated polyester fibers have greater ease in factory production and installation; lower cost; and medium durability—around 10 years.

PTFE-coated glass fiber membranes have superior durability—around 30 years; and greater resistance to the elements (sun, rain, and winds); however, they require skilled labor.

In this system, there are two types of support: direct and indirect. The direct supports are those in which the construction is arranged directly on the rest of the building structure, while the second case is arranged from a raised point such as a mast.

The cables, which are responsible for the distribution of the tensile stresses and the hardening of the canvasses, are classified in one of two ways according to the action which they perform: load-bearing and stabilizing. Both types of cable cross orthogonally, ensuring strength in two directions and avoiding deformations. The load-bearing cables are those that directly receive the external loads, fixed at the highest points. On the other hand, the stabilizing cables are responsible for strengthening the load-bearing cables and cross the load-bearing cables orthogonally. It is possible to avoid attaching the stabilizing cables to the ground by using a peripheral fixation cable.

Further, some nomenclatures for different cables are generated according to their position: a ridge-line cable refers to the uppermost cable; while valley cables are fixed below all other cables; radial cables are stabilizer cables in the form of a ring. Ridge-line cables support gravitational loads while valley cables support wind loads.

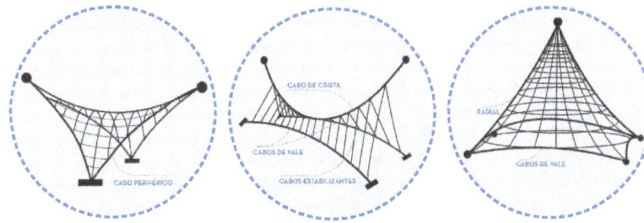

## Truss

A truss is essentially a triangulated system of straight interconnected structural elements. The most common use of trusses is in buildings, where support to roofs, the floors and internal loading such as services and suspended ceilings, are readily provided. The main reasons for using trusses are:

- Long span.

- Lightweight.

- Reduced deflection (compared to plain members).

- Opportunity to support considerable loads.

Long-span, curved roof trusses Robin Hood Airport, Doncaster

The penalty, however, is increased fabrication costs .

The article describes alternative forms of truss, where and why different forms might be appropriate and introduces design considerations. Primarily, pin jointed trusses are discussed, with some discussion of rigid-jointed Vierendeel trusses.

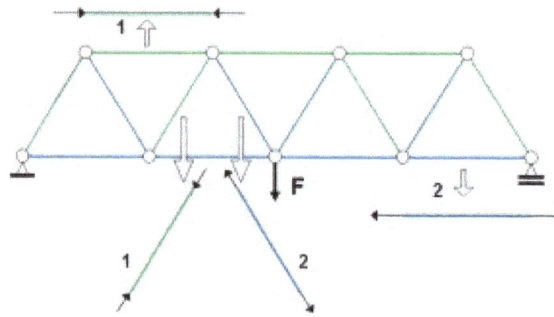

Members under axial forces in a simple truss
1 - Compression axial force
2 - Tension axial force

A truss is essentially a triangulated system of (usually) straight interconnected structural elements; it is sometimes also referred to as an open web girder. The individual elements are connected at nodes; the connections are often assumed to be nominally pinned. The external forces applied to the system and the reactions at the supports are generally applied at the nodes. When all the members and applied forces are in a same plane, the system is a plane or 2D truss.

The principal force in each element in a truss is axial tension or compression.

## Use of Trusses in Buildings

Trusses are used in a broad range of buildings, mainly where there is a requirement for very long spans, such as in airport terminals, aircraft hangers, sports stadia roofs, auditoriums and other leisure buildings. Trusses are also used to carry heavy loads and are sometimes used as transfer structures. This article focuses on typical single storey industrial buildings, where trusses are widely used to serve two main functions:

- To carry the roof load.

- To provide horizontal stability.

Two types of general arrangement of the structure of a typical single storey building are shown in the figure below.

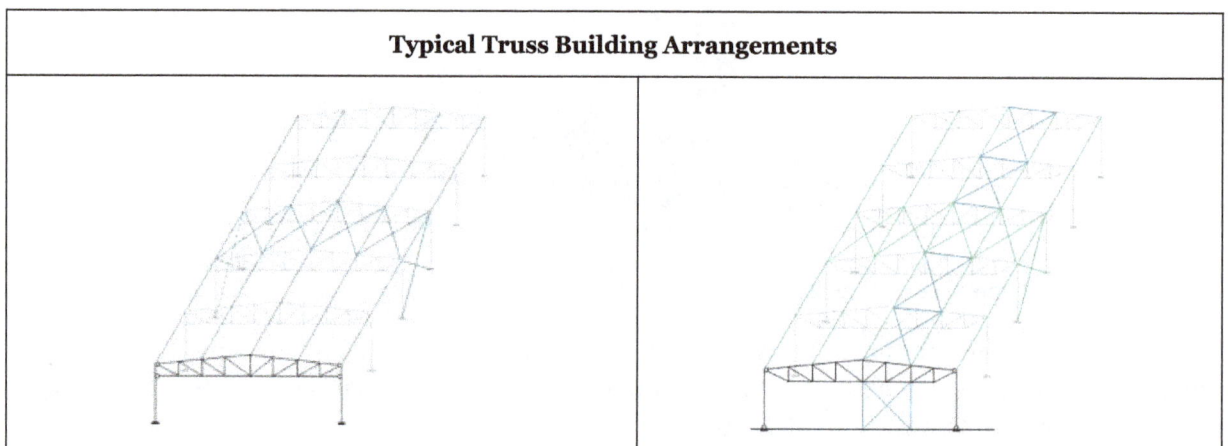

**Typical Truss Building Arrangements**

| Lateral stability provided by portal trusses. | Building braced in both directions. |
|---|---|
| Longitudinal stability provided by transverse wind girder and vertical cross bracings (blue) | Lateral stability provided by longitudinal wind girder and vertical bracings in the gables (blue) |
| No longitudinal wind girder. | Longitudinal stability provided by transverse wind girder and vertical bracings (green) |

In the first case the lateral stability of the structure is provided by a series of portal trusses; the connections between the truss and the columns provide resistance to a global bending moment. Loads are applied to the portal structure by purlins and side rails.

In the second case, each truss and the two columns between which it spans, constitute a simple structure; the connection between the truss and a column does not resist the global bending moment, and the two column bases are pinned. Bracing in both directions is necessary at the top level of the simple structure; it is achieved by means of a longitudinal wind girder which carries the transverse forces due to wind on the side walls to the vertical bracing in the gable walls. Longitudinal stability is also provided by a wind girder in the roof and vertical bracing in the elevations.

## Types of Trusses

Trusses comprise assemblies of tension and compression elements. Under gravity loads, the top and bottom chords of the truss provide the compression and tension resistance to overall bending, and the bracing resists the shear forces. A wide range of truss forms can be created. Each can vary in overall geometry and in the choice of the individual elements. Some of the commonly used types are shown below.

## Pratt Truss 'N' Truss

Pratt trusses are commonly used in long span buildings ranging from 20 to 100 m in span. In a conventional Pratt truss, diagonal members are in tension for gravity loads. This type of truss is used where gravity loads are predominant. An alternative Pratt truss is shown (below right) where the diagonal members are in tension for uplift loads. This type of truss is used where uplift loads are predominant, which may be the case in open buildings such as aircraft hangers.

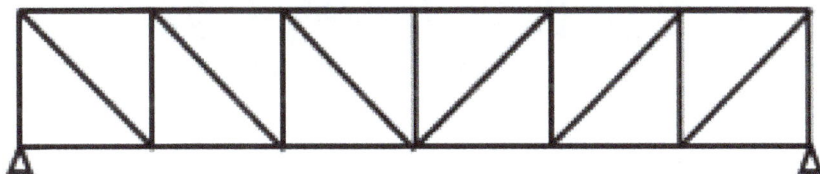

Pratt truss (gravity loads)

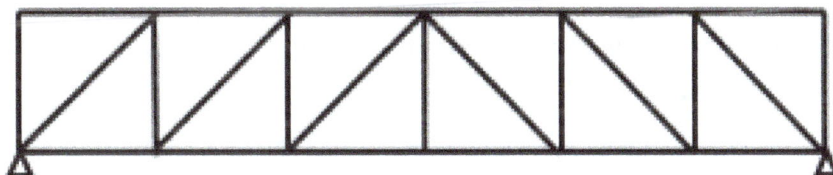

Pratt truss (uplift loads)

It is possible to add secondary members (as illustrated below left) to:

- Create intermediate support points for applied loads.

- Limit the buckling length of members in compression (although in a 2D truss, the buckling length is only modified in one axis).

For the Pratt truss and any of the types of truss mentioned below, it is possible to provide either a single or a double slope to the upper chord of a roof supporting truss. An example of a double (duo-pitch) Pratt truss is shown below.

Pratt truss with secondary members

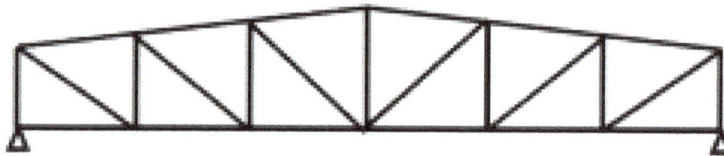

Duo-pitch Pratt truss

A Pratt truss – University of Manchester

## Warren Truss

In this type of truss, diagonal members are alternatively in tension and in compression. The Warren truss has equal length compression and tension web members, and fewer members than a Pratt truss. A modified Warren truss may be adopted where additional members are introduced to provide a node at (for example) purlin locations.

Warren trusses are commonly used in long span buildings ranging from 20 to 100 m in span.

This type of truss is also used for the horizontal truss of gantry/crane girders.

Modified Warren trusses – National Composites Centre, Bristol

Modified Warren truss

## North Light Truss

North light trusses are traditionally used for short spans in industrial workshop-type buildings. They allow maximum benefit to be gained from natural lighting by the use of glazing on the steeper pitch which generally faces north or north-east to reduce solar gain. On the steeper sloping portion of the truss, it is typical to have a truss running perpendicular to the plane of the North Light truss, to provide large column-free spaces.

The use of north lights to increase natural day lighting can reduce the operational carbon emissions of buildings although their impact should be explored using dynamic thermal modelling. Although north lights reduce the requirement for artificial lighting and can reduce the risk of overheating, by increasing the volume of the building they can also increase the demand for space heating. Further guidance is given in the Target Zero Warehouse buildings design guide.

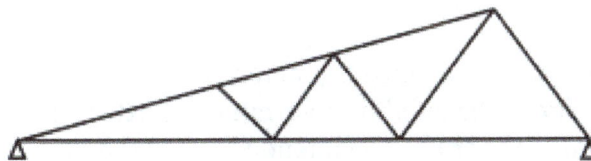

North Light truss

## Saw-tooth Truss

A variation of the North light truss is the saw-tooth truss which is used in multi-bay buildings. Similar to the North light truss, it is typical to include a truss of the vertical face running perpendicular to the plane of the saw-tooth truss.

Saw-tooth (or Butterfly) truss.

# Fink Truss

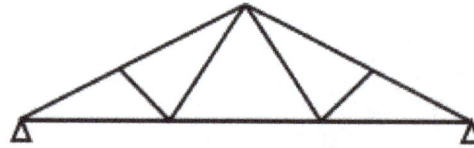

Fink truss

The Fink truss offers economy in terms of steel weight for short-span high-pitched roofs as the members are subdivided into shorter elements. There are many ways of arranging and subdividing the chords and internal members.

This type of truss is commonly used to construct roofs in houses.

## Aspects of Truss Design for Roofs

### Truss or I Beam

For the same steel weight, it is possible to get better performance in terms of resistance and stiffness, with a truss than an I beam. This difference is more sensitive for long spans and/or heavy loads. The full use of this advantage is achievable if the height of the truss is not limited by criteria other than the structural efficiency, e.g. a limit on total height of the building. However, fabrication of a truss is generally more time consuming than for an I beam, even considering that modern fabrication equipment is highly efficient.

The balance between minimum weight and minimum cost depends on many conditions: the equipment of the fabrication factory, the local cost of manufacturing; the steel unit cost, etc. Trusses generally give an economic solution for spans over 20 m.

An advantage of the truss design for roofs is that ducts and pipes that are required for operation of the buildings services can be installed through the truss web, i.e. service integration.

## General Geometry

For efficient structural performance, the ratio of span to truss depth should be chosen in the range 10 to 15. The architectural design of the building determines its external geometry and governs the slope(s) given to the top chord of the truss. The intended use of the internal space can lead either to the choice of a horizontal bottom chord, e.g. where conveyors must be hung under the chord, or to an inclined bottom chord, to allow maximum space to be provided.

For an efficient layout of the truss members between the chords, the following is advisable:

- The inclination of the diagonal members in relation to the chords should be between 35° and 55°.

- Point loads should only be applied at nodes.

- The orientation of the diagonal members should be such that the longest members are subject to tension (the shorter ones being subject to compression).

## Types of Truss Member Sections

Many solutions are available. Choice of members depends on the magnitude of the internal forces, ease of connections between members, aesthetics and any necessity to connect prefabricated truss sections on site. When selecting members, the out-of-plane buckling resistance will be important, together with resistance under reversed loading, for example, uplift.

Typical element cross sections for light building trusses

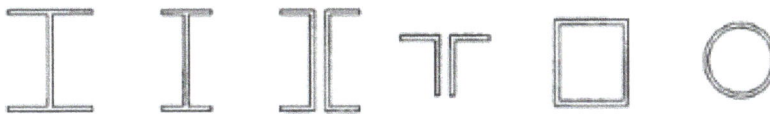

Different types of steel section used in trusses

For smaller spans, tee sections are frequently used for chords, with angles used as internal members. The internal members may be bolted or welded to the tees. Back-to-back angles or channels may be used for longer spans or heavier loads, with a gusset plate used at nodes to connect the members.

For large trusses and heavy loads, typically found in transfer trusses in buildings, members may be rolled sections; typically UKC sections. Nodes are usually welded. Any necessary connections are completed with bolted splices within the length between nodes.

Bolted angles to form lightweight, long-span trusses

For many exposed trusses, hollow sections are chosen for their structural efficiency and for aesthetic reasons. Nodes will generally be welded in the workshop. As part of the truss design, it is essential to verify the resistance of the joints (in accordance with BS EN 1993-1-8) as the joint design may dominate member selection and final truss geometry. Members should be selected carefully to avoid expensive strengthening of trusses fabricated from hollow sections.

## Types of Connections

For all the types of member sections, it is possible to design either bolted or welded connections. Generally in steelwork construction, bolted site splices are preferred to welded splices for economy and speed of erection. Where bolted connections are used, it is necessary to evaluate the consequences of 'slack' in connections. In order to reduce these consequences (typically, the increase of the deflections), pre-loaded assemblies to produce non-slip joints are recommended.

Hollow sections are typically connected by welding whilst open sections are connected by bolting or welding, which will usually involve the use of gusset plates. Guidance on the design of welded joints for Celsius®355 and Hybox®355 hollow sections is available from Tata Steel.

Small trusses which can be transported whole from the fabrication factory to the site, can be entirely welded. In the case of large roof trusses which cannot be transported whole, welded sub-assemblies are delivered to site and are either bolted or welded together on site.

In light roof trusses, entirely bolted connections are less favoured than welded connections due to the requirement for gusset plates and their increased fabrication costs.

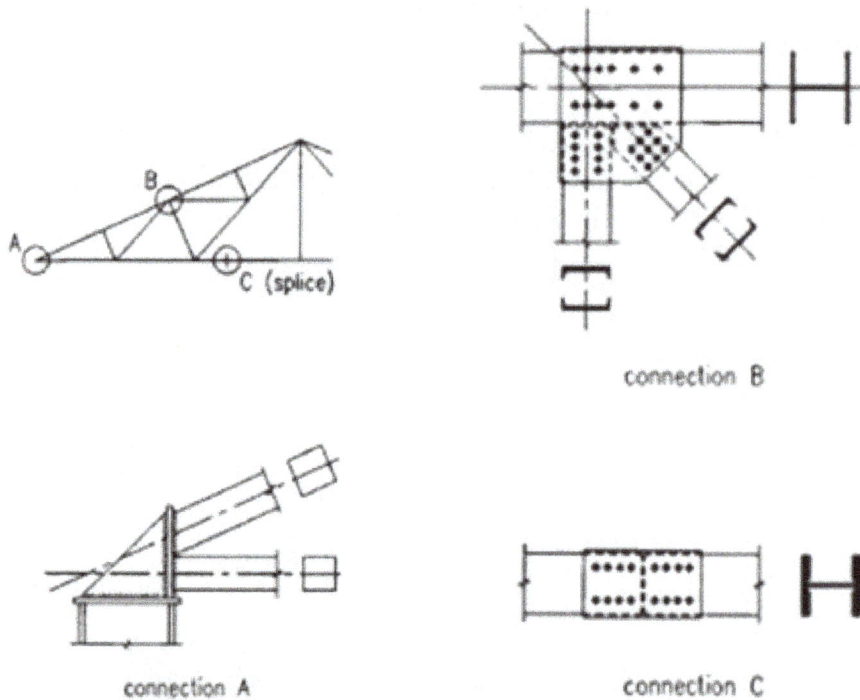

connection B

connection A                              connection C

Typical joints in welded building roof trusses

## Lateral Stability

It is necessary to design members in compression against out-of-plane buckling. For simply supported trusses, the upper chord is in compression for gravity loading, and the bottom chord is in compression for uplift loading. For portal trusses, each chord is partly in compression and partly in tension.

Lateral restraint of the upper chord is generally given by the purlins and the transverse roof wind girder.

For the restraint of the bottom chord, additional bracing may be necessary, as shown below. Such bracing allows the buckling length of the bottom chord to be limited out of the plane of the truss to the distance between points laterally restrained; the diagonal members transfer the restraint forces to the level of the top chord, where the general roof bracing is provided.

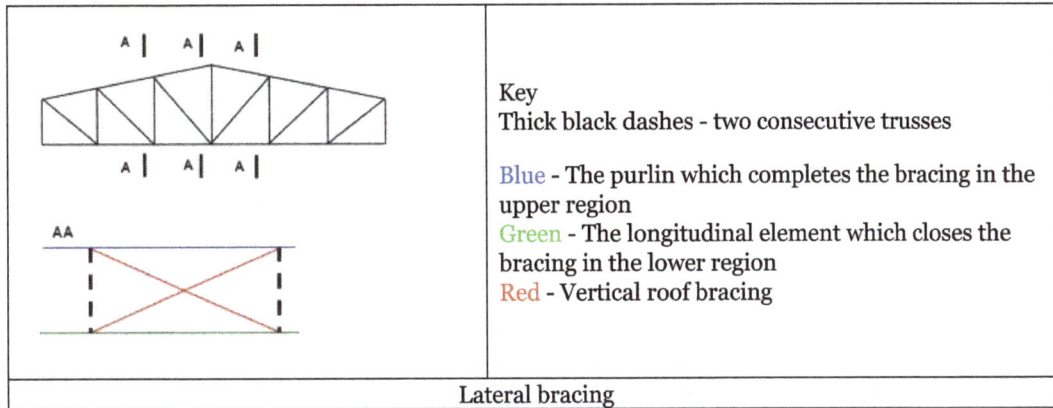

Key
Thick black dashes - two consecutive trusses

Blue - The purlin which completes the bracing in the upper region
Green - The longitudinal element which closes the bracing in the lower region
Red - Vertical roof bracing

Lateral bracing

It is possible to create a horizontal wind girder at the level of the bottom chords, with longitudinal elements to stabilize all the trusses.

## Design of Wind Girders

### Transverse Wind Girder

In general, the form of a transverse wind girder is as follows:

- The wind girder is arranged as a Warren or Pratt truss, parallel to the roof plane.

- The chords of the wind girder are the upper chords of two adjacent vertical trusses. This means that the axial forces in these members due to loading on the vertical truss and those due to loads on the wind girder loading must be added together (for an appropriate combination of actions).

It is convenient to arrange a transverse wind girder at each end of the building so that the longitudinal members need act only in tension.

### Longitudinal Wind Girder

It is necessary to provide a longitudinal wind girder (between braced gable ends) in buildings where the roof trusses are not 'portalized'.

The general arrangement is similar to that described for a transverse wind girder:

- Warren or Pratt truss.

- Generally, chord members will be provided from hollow sections.

- The posts (if required) are the upper chords of the consecutive stabilized roof trusses.

## Guidance on Global Analysis

Although joints in trusses are often hardly pinned in reality, it is generally satisfactory (and encouraged by design standards) to assume the joints are pinned and to verify the members for axial load only.

If loads are applied between nodes, trusses are often analysed with continuous chords, but with all internal members pinned. These assumptions about pinned joint behaviour apply to both bolted and welded connections.

Where member centre lines do not intersect at a node (the joint geometry may have been adjusted to increase the strength of the joint), the additional moments produced by the eccentricity are usually allowed for in the design of the chord members.

## Modelling

Several questions arise in respect of the modelling of a truss.

It is usually convenient to work on restricted models. For example, for a standard building, it is common and usually justified to work with 2D models (portal, wind girder, vertical bracing) rather than a global 3D model. A truss can be modelled without its supporting columns when it is articulated to the columns.

Nonetheless, it is important to note that:

- If separate models are used, it may be necessary, in order to verify the resistance of certain elements, to combine the results of several analyses; example: the upper chord of a truss also serves as chord of the wind girder.

- If a global 3D model is used and appropriate member releases not provided, 'parasitic' bending can be observed, which often only creates an illusory precision of the structural behaviour process.

For trusses, two analysis models are commonly used, either:

- Continuous chords with pinned internals, or

- Pinned joints throughout the truss.

## The Effect of Non-preloaded Assemblies on Truss Deflection

When the connections between elements which make up a truss are bolted, with bolts in shear and bearing (category A in BS EN 1993-1-8 ), the clearance introduced into these connections (which allows slip) can have a significant effect on displacement of the nodes.

In order to facilitate erection, the bolts are located in holes which are larger than the bolts themselves. For standard bolt sizes, holes which are 2 mm bigger than the bolt are usually made (usually referred to as a clearance hole).

Tubular trusses as an aesthetic feature in a single storey building

In order for a connection with clearance holes to transmit the load, the bolt must come into contact with one or other of the connected parts which allows slip in the connection. For a connected tension member, this slip can be considered as an additional extension that is added to the elastic elongation of the member in tension. Likewise, for a connected compression member, the slip is considered as a reduction in length that is added to the elastic shortening of the compressed member.

The total slip in the many different connections of a truss structure can lead to a significant increase in displacements, which can have more or less serious consequences:

- In most of the cases, the visual effect is the worst consequence.

- Increased deflection can lead to a reduction of free height under the bottom chord, which might prevent or upset the anticipated usage. For example, the additional deflection of a truss holding doors suspended in a gable of an aeroplane hangar could prevent the passage of the aeroplane.

- Increase in the deflection can result in reduction in the slope of the supported roof and even, if the nominal slope were small, to a slope inversion; the risk of water ingress is increased.

It is therefore essential, where truss structures are concerned, to control the effect of connection slack on the displacements. In order to do this, it is often necessary:

- To use preloaded bolts (category C connections).

- To use welded connections instead of bolted connections.

## Detailed Design Considerations for Elements

Truss members are subjected to axial force, but may also be subjected to bending moments, for example, if the chords have been modelled as continuous.

## Verification of Members Under Compression

The resistance of a member to compression is evaluated by taking into account the different modes of instability:

- Local buckling of the section is controlled using section classification.

- Buckling of the member is controlled by applying a reduction factor to the resistance of the cross-section.

In most truss members, only flexural buckling of the compressed members in the plane of the truss structure and out of the plane of the truss structure need be evaluated.

The buckling resistance is obtained from BS EN 1993-1-1 by applying a reduction to the resistance of the cross-section. This reduction factor is obtained from the slenderness of the member, which depends on the elastic critical force.

For the diagonals and the verticals stressed in uniform compression the elastic critical force is determined from the buckling length of the member in accordance with BS EN 1993-1-1 Section 6.3.1.3 and according to Annex BB of BS EN 1993-1-1:

- For buckling in the plane of the truss, the buckling length is taken equal to 90% of the system length (distance between nodes), when the truss member is connected at each end with at least two bolts, or by welding.

- For buckling out of plane of the truss beam, the buckling length is taken equal to the system length.

For buckling in the plane of the truss of the chord members in uniform compression, the buckling length may be taken as 90% of its system length (distance between nodes).

For buckling out of plane of the truss, the buckling length must be taken between lateral support points.

In the worked example, where the truss supports a roof, with purlins at the level of the upper chord of the truss:

- All the purlins connected to a roof bracing can be considered as lateral rigid support points.

- Intermediate purlins can also be considered as a rigid point of support, if the roof behaves as a diaphragm (class 2 construction according to BS EN 1993-1-3).

- Lateral support points are provided to the lower chord by additional vertical bracing elements between trusses.

## Vierendeel Trusses

## Use of Vierendeel Trusses

Vierendeel trusses are rigidly-jointed trusses having only vertical members between the top and bottom chords. The chords are normally parallel or near parallel.

Elements in Vierendeel trusses are subjected to bending, axial force and shear , unlike conventional trusses with diagonal web members where the members are primarily designed for axial loads.

Vierendeel trusses are usually more expensive than conventional trusses and their use limited to instances where diagonal web members are either obtrusive or undesirable.

Vierendeel trusses are moment resisting. Vertical members near the supports are subject to the highest moments and therefore require larger sections to be used than those at mid-span. Considerable bending moments must be transferred between the verticals and the chords, which can result in expensive stiffened details.

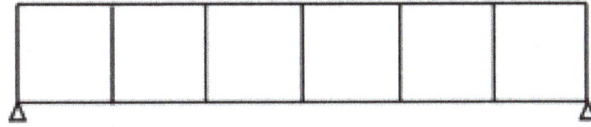

Vierendeel truss

## Analysis

As Vierendeel trusses are statically indeterminate structures, computer analysis software packages are generally used to analyse the truss.

## Staggered Truss System

In an effort to provide a structural-steel framing system with a minimum floor-to- floor height for multi-storey residential construction, the staggered truss system was developed. This system consists of story-high trusses spanning the full width of a building. They are placed at alternate column lines in alternate stories, thus resulting in a staggered arrangement of trusses. The trusses span about 60 ft. between exterior columns, resulting in a column free interior space. In addition to the simple checkerboard pattern, alternative stacking patterns are possible in order to accommodate varied interior layouts.

At a typical floor, the deck spans between the top chord of one truss and the bottom chord of the adjacent truss. Since the staggered trusses are typically spaced 20 to 30 ft. on centres, a long-span floor deck system is required. Precast-concrete plank with topping is frequently used, since, in addition to accommodating the span the plank underside can be finished to provide an acceptable ceiling. An alternative system consists of long-span composite metal deck, having a depth of up to 7½ in, with concrete fill. The top and bottom chords of the trusses are usually wide-flange shapes to efficiently resist the stresses induced by the floor loadings.

Diagonal web members of the trusses are deleted at corridor openings. This results in bending stresses in the truss chords due to Vierendeel action. Consequently, corridors are typically located near the building centreline that is, near midspan of the trusses, at points of minimum truss shear, thereby minimizing the chord bending stresses.

Lateral loads in the transverse direction are transferred to the truss top chords via diaphragm action of the floor deck. These loads are transmitted through the depth of the trusses to the bottom chords and are then transferred through the floor deck at that level to the adjacent-truss top chords. The overturning couple produced by the transfer of lateral load from the top chord to the bottom chord is resisted by a vertical couple at the ends of the truss. Only axial forces are induced in the exterior columns. Therefore, transverse lateral loads are transmitted down through the structure without creating bending stresses in the trusses or columns, except at truss openings.

In the longitudinal direction, lateral loads are transferred via floor diaphragm action to the exterior columns. These resist the loads by conventional means, such as rigid frames or braced bents. To provide added strength and stiffness, the exterior columns are usually oriented so that the strong axis assists in resisting lateral loads in the longitudinal direction.

To achieve the necessary structural interaction between the trusses and the floor deck and to provide the necessary continuity of the floor diaphragm, adequate connection by such means as weld plates or shear connectors must be provided between the various structural elements. Floor decks with large openings or other shear discontinuities may require additional reinforcement.

Although the staggered-truss system resists gravity and lateral loads primarily by axial stresses, consideration must be given to the bending stresses in the exterior columns that result from the truss deformations under gravity loads. These bending stresses can be significantly reduced by cambering the trusses, thereby preloading the columns. An alternative is to provide slotted bottom-chord connections that are torqued or welded after dead load is applied.

Its main advantages include reduced floor-to-floor height, 20-25% faster construction time, more column-free space, and a lighter overall structural system.

## References

- Design-of-steel-compression-members-4896: theconstructor.org, Retrieved 16 May 2018

- What-is-beam-types-of-beams: mech4study.com, Retrieved 11 July 2018

- What-are-the-types-of-ties-used-in-building-construction-19000: theconstructor.org, Retrieved 31 June 2018

- What-are-the-building-structural-systems: gharpedia.com, Retrieved 19 April 2018

- What-is-a-rigid-frame-building-378452: whirlwindsteel.com, Retrieved 29 June 2018

- Analysis-space-frame-structure: masterbuilder.co.in, Retrieved 10 July 2018

- Tensile-structures-how-do-they-work-and-what-are-the-different-types-887462: archdaily.com, Retrieved 19 March 2018

# Structural Materials

# 3

A study of various building materials and their properties is vital to the understanding of how these materials support structures and resist loads. Some of the common structural materials are con-crete, iron, timber, alloys, aluminium, composite materials, etc. This chapter has been carefully written to provide an extensive understanding of common structural materials.

Structural materials are materials used or studied primarily for their mechanical properties, as opposed to their electronic, magnetic, chemical or optical characteristics. This can include a materials response to an applied force, whether this response is elastic or plastic, its hardness, and its strength.

In ideal environments, most common construction materials are very durable and can last indefinitely. However, design or construction deficiencies or lack of proper maintenance can result in less-than-ideal conditions under which construction materials will degrade. Degradation can take many forms, including chemical reactions, consumption by living organisms, and erosion or mechanical wear. Traditional building materials – steel, concrete, and wood – usually deteriorate and fail via well-known mechanisms. Even innovative materials that appear on construction sites can degrade, either by these well-understood mechanisms or through exotic, sometimes surprising, reactions and processes.

Brick is the oldest of all artificial building materials. It is classified as face brick, common brick, and glazed brick. Face brick isused on the exterior of a wall and varies in color, texture, and mechanical perfection. Common brick consists of the kiln run ofbrick and is used behind whatever facing material is employed providing necessary wall thickness and additional structuralstrength. Glazed brick is employed largely for interiors where beauty, ease of cleaning, and sanitation are primaryconsiderations.

Structural clay tiles are burned-clay masonry units having interior hollow spaces termed cells. Such tile is widely usedbecause of its strength, light weight, and insulating and fire-protection qualities.

Architectural terra-cotta is a burned-clay material used for decorative purposes. The shapes are molded either by hand inplaster-of-paris molds or by machine, using the stiff-mud process.

Building stones generally used are limestone, sandstone, granite, and marble. Until the advent of steel and concrete, stonewas the most important building material. Its principal use now is as a decorative material because of its beauty, dignity, anddurability.

Concrete is a mixture of cement, mineral aggregate, and water, which, if combined in proper proportions, form a plasticmixture capable of being placed in forms and of hardening through the hydration of the cement.

The cellular structure of wood is largely responsible for its basic characteristics, unique among the common structuralmaterials. When cut into lumber, a tree provides a wide range of material which is classified according to use as yard lumber,factory or shop lumber, and structural lumber. Laminated lumber is used for beams, columns, arch ribs, chord members, andother structural members. Plywood is generally used as a replacement for sheathing, or as form lumber for reinforced concrete structures.

Important structural metals are the structural steels, steel castings, aluminum alloys, magnesium alloys, and cast andwrought iron. Steel castings are used for rocker bearings under the ends of large bridges. Shoes and bearing plates areusually cast in carbon steel, but rollers are often cast in stainless steel. Aluminum alloys are strong, lightweight, and resistantto corrosion. The alloys most frequently used are comparable with the structural steels in strength. Magnesium alloys areproduced as extruded shapes, rolled plate, and forgings. The principal structural applications are in aircraft, truck bodies, andportable scaffolding. Gray cast iron is used as a structural material for columns and column bases, bearing plates, stairtreads, and railings. Malleable cast iron has few structural applications. Wrought iron is used extensively because of its abilityto resist corrosion. It is used for blast plates to protect bridges, for solid decks to support ballasted roadways, and for trash racks for dams.

Composite materials are engineered materials that contain a load-bearing material housed in a relatively weak protectivematrix. A composite material results when two or more materials, each having its own, usually different characteristics, arecombined, producing a material with properties superior to its components. The matrix material (metallic, ceramic, orpolymeric) bonds together the reinforcing materials (whiskers, laminated fibers, or woven fabric) and distributes the loadingbetween them.

Fiber-reinforced polymers (FRP) are a broad group of composite materials made of fibers embedded in a polymeric matrix.Compared to metals, they generally have relatively high strength-to-weight ratios and excellent corrosion resistance. Theycan be formed into virtually any shape and size. Glass is by far the most used fiber in FRP (glass-FRP), although carbonfiber (carbon-FRP) is finding greater application. Although complete FRP shapes and structures are possible, the mostpromising application of FRP in civil engineering is for repairing structures or infrastructure. FRP can be used to repairbeams, walls, slabs, and columns.

## Iron

Iron is the fourth most common element in Earth's crust (after oxygen, silicon, and aluminum), and the second most common metal (after aluminium), but because it reacts so readily with oxygen it's never mined in its pure form (though meteorites are occasionally discovered that contain samples of pure iron). Like aluminium, most iron "locked" inside Earth exists in the form of oxides (compounds of iron and oxygen). Iron oxides exist in seven main ores (raw, rocky minerals mined from Earth):

- Hematite (the most plentiful).

- Limonite (also called brown ore or bog iron).

- Goethite.

- Magnetite (black ore; the magnetic type of iron oxide, also called lodestone).

- Pyrite.

- Siderite.

- Taconite (a combination of hematite and magnetite).

Different ores contain different amounts of iron. Hematite and magnetite have about 70 percent iron, limonite has about 60 percent, pyrite and siderite have 50 percent, while taconite has only 30 percent. Using a combination of both deep mining (under the ground) and opencast mining (on the surface), the world produces approximately 1000 million tons of iron ore each year, with China responsible for just over half of it.

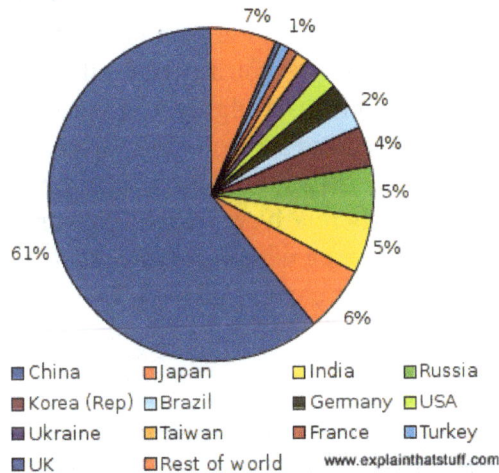

*Which countries produce the world's iron? Chart shows estimated figures for pig iron for. In the United States, three companies currently produce pig iron in nine different locations.*

## Types of Iron

Pure iron is too soft and reactive to be of much real use, so most of the "iron" we tend to use for everyday purposes is actually in the form of iron alloys: iron mixed with other elements (especially carbon) to make stronger, more resilient forms of the metal including steel. Broadly speaking, steel is an alloy of iron that contains up to about 2 percent carbon, while other forms of iron contain about 2–4 percent carbon. In fact, there are thousands of different kinds of iron and steel, all containing slightly different amounts of other alloying elements.

## Pig Iron

Basic raw iron is called pig iron because it's produced in the form of chunky molded blocks known as pigs. Pig iron is made by heating an iron ore (rich in iron oxide) in a blast furnace: an enormous industrial fireplace, shaped like a cylinder, into which huge drafts of hot air are introduced in regular "blasts". Blast furnaces are often spectacularly huge: some are 30–60m (100–200ft) high, hold dozens of trucks worth of raw materials, and often operate continuously for years at a time without being switched off or cooled down. Inside the furnace, the iron ore reacts chemically with coke (a carbon-rich form of coal) and limestone. The coke "steals" the oxygen from the iron oxide (in a chemical process called reduction), leaving behind a relatively pure liquid iron, while the limestone helps to remove the other parts of the rocky ore (including clay, sand, and small stones), which form a waste slurry known as slag. The iron made in a blast furnace is an alloy containing about 90–95 percent iron, 3–4 percent carbon, and traces of other elements such as silicon, manganese, and phosphorus, depending on the ore used. Pig iron is much harder than 100 percent pure iron, but still too weak for most everyday purposes.

## Cast Iron

One of the world's most famous iron buildings, the Capitol in Washington, DC has a dome made of 4,041,146kg (8,909,200 pounds) of cast iron. Cast iron is simply liquid iron that has been cast: poured into a mold and allowed to cool and harden to form a finished structural shape, such as a pipe, a gear, or a big girder for an iron bridge. Pig iron is actually a very basic form of cast iron, but it's molded only very crudely because it's typically melted down to make steel. The high carbon content of cast iron (the same as pig iron—roughly 3–4 percent) makes it extremely hard and brittle: large crystals of carbon embedded in cast iron stop the crystals of iron from moving about. Cast iron has two big drawbacks: first, because it's hard and brittle, it's virtually impossible to shape, even when heated; second, it rusts relatively easily. It's worth noting that there are actually several different types of cast iron, including white and gray cast irons (named for the coloring of the finished product caused by the way the carbon inside it behaves).

## Wrought Iron

Cast iron assumes its finished shape the moment the liquid iron alloy cools down in the mold. Wrought iron is a very different material made by mixing liquid iron with some slag (leftover waste). The result is an iron alloy with much lower carbon content. Wrought iron is softer than cast iron and much less tough, so you can heat it up to shape it relatively easily, and it's also much less prone to rusting. However, relatively little wrought iron is now produced commercially, since most of the objects originally produced from it are now made from steel, which is both cheaper and generally of more consistent quality. Wrought iron is what people used to use before they really mastered making steel in large quantities in the mid-19th century.

Left: Pig iron is the raw material used to make other forms of iron and steel. Middle: Cast iron was used for strong, structural components like bits of engines and bridges. Right: Wrought iron is a softer iron still widely used to make everyday things like these street railings.

# Characteristics

## Mechanical Properties

| Characteristic values of tensile strength(TS) and Brinell hardness (BH) of different forms of iron. | | |
|---|---|---|
| **Material** | **TS (MPa)** | **BH (Brinell)** |
| Iron whiskers | 11000 | |
| Ausformed (hardened) steel | 2930 | 850–1200 |
| Martensitic steel | 2070 | 600 |
| Bainitic steel | 1380 | 400 |
| Pearlitic steel | 1200 | 350 |
| Cold-worked iron | 690 | 200 |
| Small-grain iron | 340 | 100 |
| Carbon-containing iron | 140 | 40 |
| Pure, single-crystal iron | 10 | 3 |

The mechanical properties of iron and its alloys can be evaluated using a variety of tests, including the Brinell test, Rockwell test and the Vickers hardness test. The data on iron is so consistent that it is often used to calibrate measurements or to compare tests. However, the mechanical properties of iron are significantly affected by the sample's purity: pure, single crystals of iron are actually softer than aluminium, and the purest industrially produced iron (99.99%) has a hardness of 20–30 Brinell. An increase in the carbon content will cause a significant increase in the hardness and tensile strength of iron. Maximum hardness of 65 $R_c$ is achieved with a 0.6% carbon content, although the alloy has low tensile strength. Because of the softness of iron, it is much easier to work with than its heavier congeners ruthenium and osmium.

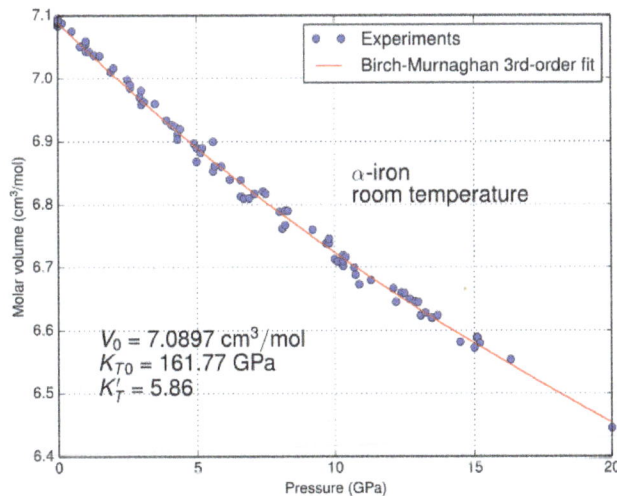

Molar volume vs. pressure for α iron at room temperature

Because of its significance for planetary cores, the physical properties of iron at high pressures and temperatures have also been studied extensively. The form of iron that is stable under standard conditions can be subjected to pressures up to ca. 15 GPa before transforming into a high-pressure form.

# Phase Diagram and Allotropes

Iron represents an example of allotropy in a metal. At least four allotropic forms of iron are known as α, γ, δ, and ε; at very high pressures and temperatures, some controversial experimental evidence exists for a stable β phase.

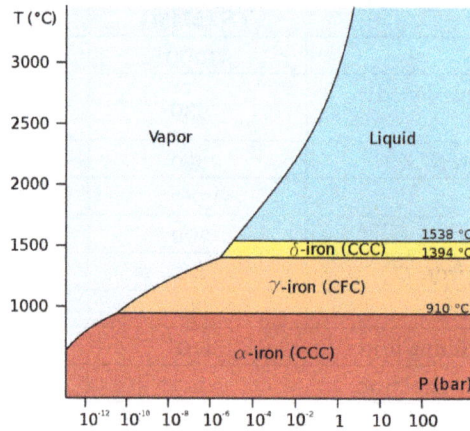

Low-pressure phase diagram of pure iron

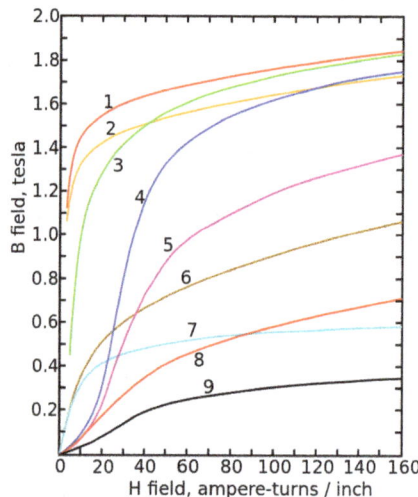

Magnetization curves of 9 ferromagnetic materials, showing saturation.
1. Sheet steel, 2. Silicon steel, 3. Cast steel, 4. Tungsten steel, 5. Magnet steel, 6. Cast iron, 7. Nickel, 8. Cobalt, 9. Magnetite

As molten iron cools past its freezing point of 1538 °C, it crystallizes into its δ allotrope, which has a body-cantered cubic (bcc) crystal structure. As it cools further to 1394 °C, it changes to its γ-iron allotrope, a face-centered cubic (fcc) crystal structure, or austenite. At 912 °C and below, the crystal structure again becomes the bcc α-iron allotrope. Finally, at 770 °C (the Curie point, $T_c$) iron's magnetic ordering changes from paramagnetic to ferromagnetic. As is passes through the Curie temperature, iron does not change its structure, but "magnetic domains" appear where each domain contains iron atoms with a particular electronic spin. In unmagnetized iron, all the electronic spins of the atoms within one domain have the same axis orientation; however, the electrons of neighboring domains have other orientations with the result of mutual cancellation and no magnetic field. In magnetized iron, the electronic spins of the domains are aligned and the magnetic effects are reinforced. Although each domain contains billions of atoms, they are very small, about 10 micrometres across. This happens because the two unpaired electrons on each iron atom are in the $d_z^2$ and $d_{x^2-y^2}$ orbitals, which do not point directly at the nearest neighbours in the body-cantered cubic lattice and therefore do not participate in metallic bonding; thus, they can interact magnetically with each other so that their spins align.

At pressures above approximately 10 GPa and temperatures of a few hundred kelvin or less, α-iron changes into a hexagonal close-packed (hcp) structure, which is also known as ε-iron; the higher-temperature γ-phase also changes into ε-iron, but does so at higher pressure. The β-phase, if it exists, would appear at pressures of at least 50 GPa and temperatures of at least 1500 K and have an orthorhombic or a double hcp structure. These high-pressure phases of iron are important as end member models for the solid parts of planetary cores. The inner core of the Earth is generally presumed to be an iron-nickel alloy with ε (or β) structure. Somewhat confusingly, the term "β-iron" is sometimes also used to refer to α-iron above its Curie point, when it changes from being ferromagnetic to paramagnetic, even though its crystal structure has not changed.

The melting point of iron is experimentally well defined for pressures less than 50 GPa. For greater pressures, studies put the γ-ε-liquid triple point at pressures that differ by tens of gigapascals and 1000 K in the melting point. Generally speaking, molecular dynamics computer simulations of iron melting and shock wave experiments suggest higher melting points and a much steeper slope of the melting curve than static experiments carried out in diamond anvil cells. The melting and boiling points of iron, along with its enthalpy of atomization, are lower than those of the earlier 3d elements from scandium to chromium, showing the lessened contribution of the 3d electrons to metallic bonding as they are attracted more and more into the inert core by the nucleus; however, they are higher than the values for the previous element manganese because that element has a half-filled 3d subshell and consequently its d-electrons are not easily delocalized. This same trend appears for ruthenium but not osmium.

## Isotopes

Naturally occurring iron consists of four stable isotopes: 5.845% of $^{54}$Fe, 91.754% of $^{56}$Fe, 2.119% of $^{57}$Fe and 0.282% of $^{58}$Fe. Of these stable isotopes, only $^{57}$Fe has a nuclear spin $(-\frac{1}{2})$. The nuclide $^{54}$Fe theoretically can undergo double electron capture to $^{54}$Cr, but the process has never been observed and only a lower limit on the half-life of $3.1 \times 10^{22}$ years has been established.

$^{60}$Fe is an extinct radionuclide of long half-life (2.6 million years). It is not found on Earth, but its ultimate decay product is its granddaughter, the stable nuclide $^{60}$Ni. Much of the past work on isotopic composition of iron has focused on the nucleosynthesis of $^{60}$Fe through studies of meteorites and ore formation. In the last decade, advances in mass spectrometry have allowed the detection and quantification of minute, naturally occurring variations in the ratios of the stable isotopes of iron. Much of this work is driven by the Earth and planetary science communities, although applications to biological and industrial systems are emerging.

In phases of the meteorites *Semarkona* and *Chervony Kut,* a correlation between the concentrations of $^{60}$Ni, the granddaughter of $^{60}$Fe, and the abundance of the stable iron isotopes provided evidence for the existence of $^{60}$Fe at the time of formation of the Solar System. Possibly the energy released by the decay of $^{60}$Fe, along with that released by $^{26}$Al, contributed to the remelting and differentiation of asteroids after their formation 4.6 billion years ago. The abundance of $^{60}$Ni present in extraterrestrial material may bring further insight into the origin and early history of the Solar System.

The most abundant iron isotope $^{56}$Fe is of particular interest to nuclear scientists because it represents the most common endpoint of nucleosynthesis. Since $^{56}$Ni (14 alpha particles) is easily produced from lighter nuclei in the alpha processin nuclear reactions in supernovae (see silicon burning process), it is the endpoint of fusion chains inside extremely massive stars, since addition of another alpha particle, resulting in $^{60}$Zn, requires a great deal more energy. This $^{56}$Ni, which has a half-life of about 6 days, is created in quantity in these stars, but soon decays by two successive positron emissions within supernova decay products in the supernova remnant gas cloud, first to radioactive $^{56}$Co, and then to stable $^{56}$Fe. As such, iron is the most abundant element in the core of red giants, and is the most abundant metal in iron meteorites and in the dense metal cores of planets such as Earth. It is also very common in the universe, relative to other stable metals of approximately the same atomic weight. Iron is the sixth most abundant element in the Universe, and the most common refractory element.

Although a further tiny energy gain could be extracted by synthesizing $^{62}$Ni, which has a marginally higher binding energy than $^{56}$Fe, conditions in stars are unsuitable for this process. Element production in supernovas and distribution on Earth greatly favor iron over nickel, and in any case, $^{56}$Fe still has a lower mass per nucleon than $^{62}$Ni due to its higher fraction of lighter protons. Hence, elements heavier than iron require a supernova for their formation, involving rapid neutron capture by starting $^{56}$Fe nuclei.

In the far future of the universe, assuming that proton decay does not occur, cold fusion occurring via quantum tunnelling would cause the light nuclei in ordinary matter to fuse into $^{56}$Fe nuclei. Fission and alpha-particle emission would then make heavy nuclei decay into iron, converting all stellar-mass objects to cold spheres of pure iron.

## Chemistry and Compounds

| Oxidation State | Representative Compound |
|---|---|
| −2 (d$^{10}$) | Disodium tetracarbonylferrate (Collman's reagent) |
| −1 (d$^9$) | $Fe_2(CO)_8^{2-}$ |
| 0 (d$^8$) | Iron pentacarbonyl |
| 1 (d$^7$) | Cyclopentadienyliron dicarbonyl dimer ("Fp$_2$") |
| 2 (d$^6$) | Ferrous sulfate, ferrocene |
| 3 (d$^5$) | Ferric chloride, ferrocenium tetrafluoroborate |
| 4 (d$^4$) | $Fe(diars)_2Cl_2^{2+}$ |
| 5 (d$^3$) | $FeO_4^{3-}$ |
| 6 (d$^2$) | Potassium ferrate |
| 7 (d$^1$) | [FeO$_4$]$^-$ (matrix isolation, 4K) |

Iron shows the characteristic chemical properties of the transition metals, namely the ability to form variable oxidation states differing by steps of one and a very large coordination and organometallic chemistry: indeed, it was the discovery of an iron compound, ferrocene, that revolutionalized the latter field in the 1950s. Iron is sometimes considered as a prototype for the entire block of transition metals, due to its abundance and the immense role it has played in the technological progress of humanity. Its 26 electrons are arranged in the configuration [Ar]$3d^6 4s^2$, of which the 3d and 4s electrons are relatively close in energy, and thus it can lose a variable number of electrons and there is no clear point where further ionization becomes unprofitable.

Iron forms compounds mainly in the +2 and +3 oxidation states. Traditionally, iron (II) compounds are called ferrous, and iron (III) compounds ferric. Iron also occurs in higher oxidation states, e.g. the purple potassium ferrate ($K_2 FeO_4$), which contains iron in its +6 oxidation state. Although iron (VIII) oxide ($FeO_4$) has been claimed, the report could not be reproduced and such a species (at least with iron in its +8 oxidation state) has been found to be improbable computationally. However, one form of anionic $[FeO_4]^-$ with iron in its +7 oxidation state, along with an iron(V)-peroxo isomer, has been detected by infrared spectroscopy at 4 K after cocondensation of laser-ablated Fe atoms with a mixture of $O_2$/Ar. Iron (IV) is a common intermediate in many biochemical oxidation reactions. Numerous organoiron compounds contain formal oxidation states of +1, 0, −1, or even −2. The oxidation states and other bonding properties are often assessed using the technique of Mössbauer spectroscopy. Many mixed valence compounds contain both iron (II) and iron(III) centres, such as magnetite and Prussian blue ($Fe_4(Fe[CN]_6)_3$). The latter is used as the traditional "blue" in blueprints.

Iron is the first of the transition metals that cannot reach its group oxidation state of +8, although its heavier congeners ruthenium and osmium can, with ruthenium having more difficulty than osmium. Ruthenium exhibits an aqueous cationic chemistry in its low oxidation states similar to that of iron, but osmium does not, favouring high oxidation states in which it forms anionic complexes. In the second half of the 3d transition series, vertical similarities down the groups compete with the horizontal similarities of iron with its neighbours cobalt and nickel in the periodic table, which are also ferromagnetic at room temperature and share similar chemistry. As such, iron, cobalt, and nickel are sometimes grouped together as the iron triad.

Hydrated iron(III) chloride, also known as ferric chloride

The iron compounds produced on the largest scale in industry are iron(II) sulfate ($FeSO_4 \cdot 7H_2O$) and iron(III) chloride ($FeCl_3$). The former is one of the most readily available sources of iron(II), but is less stable to aerial oxidation than Mohr's salt (($NH_4)_2 Fe(SO_4)_2 \cdot 6H_2O$). Iron(II) compounds tend to be oxidized to iron(III) compounds in the air.

Unlike many other metals, iron does not form amalgams with mercury. As a result, mercury is traded in standardized 76 pound flasks (34 kg) made of iron.

Iron is by far the most reactive element in its group; it is pyrophoric when finely divided and dissolves easily in dilute acids, giving $Fe^{2+}$. However, it does not react with concentrated nitric acid and other oxidizing acids due to the formation of an impervious oxide layer, which can nevertheless react with hydrochloric acid.

## Binary Compounds

Iron reacts with oxygen in the air to form various oxide and hydroxide compounds; the most common are iron(II,III) oxide ($Fe_3O_4$), and iron(III) oxide ($Fe_2O_3$). Iron(II) oxide also exists, though it is unstable at room temperature. Despite their names, they are actually all non-stoichiometric compounds whose compositions may vary. These oxides are the principal ores for the production of iron (see bloomery and blast furnace). They are also used in the production of ferrites, useful magnetic storage media in computers, and pigments. The best known sulfide is iron pyrite ($FeS_2$), also known as fool's gold owing to its golden luster. It is not an iron (IV) compound, but is actually an iron (II) polysulfide containing $Fe^{2+}$ and $S_2^{2-}$ ions in a distorted sodium chloride structure.

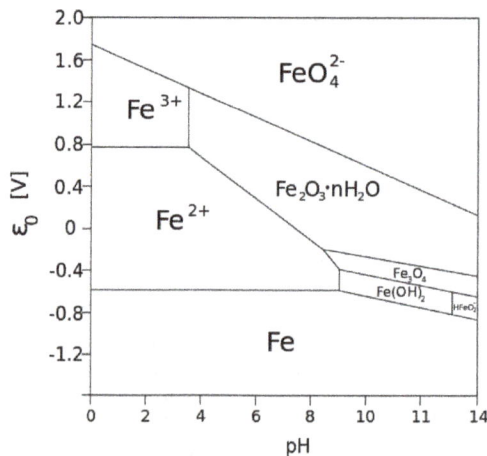

Pourbaix diagram of iron

The binary ferrous and ferric halides are well-known, with the exception of ferric iodide. The ferrous halides typically arise from treating iron metal with the corresponding hydrohalic acid to give the corresponding hydrated salts.

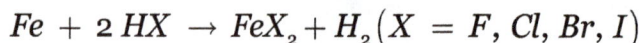

$$Fe + 2\,HX \rightarrow FeX_2 + H_2\,(X = F,\,Cl,\,Br,\,I)$$

Iron reacts with fluorine, chlorine, and bromine to give the corresponding ferric halides, ferric chloride being the most common.

$$2\,Fe + 3\,X_2 \rightarrow 2\,FeX_3\,(X = F,\,Cl,\,Br)$$

Ferric iodide is an exception, being thermodynamically unstable due to the oxidizing power of $Fe^{3+}$ and the high reducing power of $I^-$:

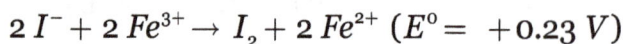

$$2\,I^- + 2\,Fe^{3+} \rightarrow I_2 + 2\,Fe^{2+}\,(E^\circ = +0.23\,V)$$

Nevertheless, milligram amounts of ferric iodide, a black solid, may still be prepared through the reaction of iron pentacarbonyl with iodine and carbon monoxide in the presence of hexane and light at the temperature of −20 °C, making sure that the system is well sealed off from air and water.

## Solution Chemistry

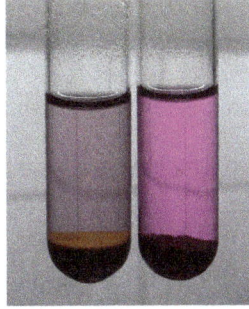

Comparison of colors of solutions of ferrate
(left) and permanganate(right)

The standard reduction potentials in acidic aqueous solution for some common iron ions are given below:

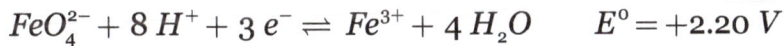

$$Fe^{2+} + 2\,e^- \quad\rightleftharpoons Fe \qquad\qquad E^\circ = -0.447\ V$$

$$Fe^{3+} + 3\,e^- \quad\rightleftharpoons Fe \qquad\qquad E^\circ = -0.037\ V$$

$$FeO_4^{2-} + 8\,H^+ + 3\,e^- \rightleftharpoons Fe^{3+} + 4\,H_2O \qquad E^\circ = +2.20\ V$$

The red-purple tetrahedral ferrate (VI) anion is such a strong oxidizing agent that it oxidizes nitrogen and ammonia at room temperature, and even water itself in acidic or neutral solutions:

$$4\,FeO_4^{2-} + 10\,H_2O \rightarrow 4\,Fe^{3+} + 20\,OH^- + 3\,O_2$$

The $Fe^{3+}$ ion has a large simple cationic chemistry, although the pale-violet hexaquo ion $[Fe(H_2O)_6]^{3+}$ is very readily hydrolysed when pH increases above 0 as follows:

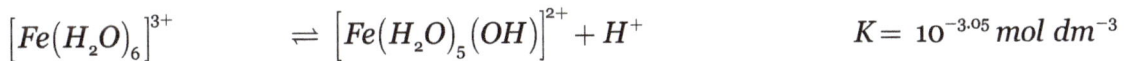

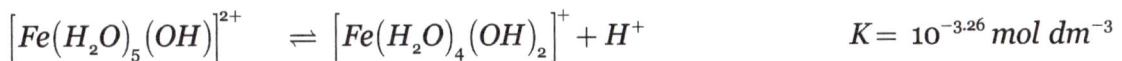

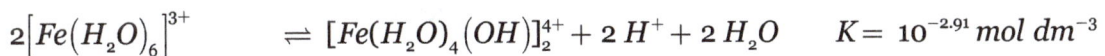

$$\left[Fe(H_2O)_6\right]^{3+} \quad\rightleftharpoons \left[Fe(H_2O)_5(OH)\right]^{2+} + H^+ \qquad\qquad K = 10^{-3.05}\ mol\ dm^{-3}$$

$$\left[Fe(H_2O)_5(OH)\right]^{2+} \rightleftharpoons \left[Fe(H_2O)_4(OH)_2\right]^{+} + H^+ \qquad\qquad K = 10^{-3.26}\ mol\ dm^{-3}$$

$$2\left[Fe(H_2O)_6\right]^{3+} \quad\rightleftharpoons [Fe(H_2O)_4(OH)]_2^{4+} + 2\,H^+ + 2\,H_2O \quad K = 10^{-2.91}\ mol\ dm^{-3}$$

Dark red iron (III) oxide

Blue-green iron (II) sulfate heptahydrate

As pH rises above 0 the above yellow hydrolysed species form and as it rises above 2–3, reddish-brown hydrous iron (III) oxide precipitates out of solution. Although $Fe^{3+}$ has an $d^5$ configuration, its absorption spectrum is not like that of $Mn^{2+}$ with its weak, spin-forbidden d–d bands, because $Fe^{3+}$ has higher positive charge and is more polarizing, lowering the energy of its ligand-to-metal charge transfer absorptions. Thus, all the above complexes are rather strongly colored, with the single exception of the hexaquo ion – and even that has a spectrum dominated by charge transfer in the near ultraviolet region. On the other hand, the pale green iron (II) hexaquo ion $[Fe(H_2O)_6]^{2+}$ does not undergo appreciable hydrolysis. Carbon dioxide is not evolved when carbonate anions are added, which instead results in white iron(II) carbonatebeing precipitated out. In excess carbon dioxide this forms the slightly soluble bicarbonate, which occurs commonly in groundwater, but it oxidises quickly in air to form iron(III) oxide that accounts for the brown deposits present in a sizeable number of streams.

## Coordination Compounds

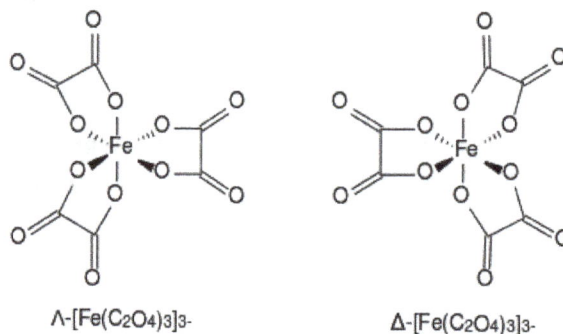

$\Lambda$-[Fe(C₂O₄)₃]³⁻                    $\Delta$-[Fe(C₂O₄)₃]³⁻

The two enantiomorphs of the ferrioxalate ion

Many coordination compounds of iron are known. A typical six-coordinate anion is hexachloroferrate(III), $[FeCl_6]^{3-}$, found in the mixed salt tetrakis (methylammonium) hexachloroferrate(III) chloride.Complexes with multiple bidentate ligands have geometric isomers. For example, the *trans*-chlorohydridobis(bis-1,2-(diphenylphosphino)ethane)iron(II) complex is used as a starting material for compounds with the Fe(dppe)$_2$ moiety. The ferrioxalate ion with three oxalate ligands (shown at right) displays helical chirality with its two non-superposable geometries labelled $\Lambda$ (lambda) for the left-handed screw axis and $\Delta$ (delta) for the right-handed screw axis, in line with IUPAC conventions. Potassium ferrioxalate is used in chemical actinometry and along with its sodium salt undergoes photoreduction applied in old-style photographic processes. The dihydrate of iron(II) oxalate has a polymeric structure with co-planar oxalate ions bridging between iron centres with the water of crystallisation located forming the caps of each octahedron, as illustrated below.

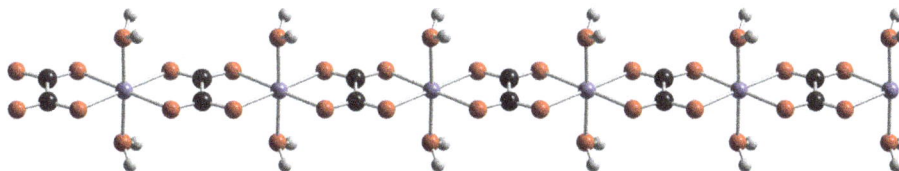

Ball-and-stick model of a chain in the crystal structure of iron(II) oxalate dihydrate

Prussian blue

Prussian blue, $Fe_4[Fe(CN)_6]_3$, is the most famous of the cyanide complexes of iron. Its formation can be used as a simple wet chemistry test to distinguish between aqueous solutions of $Fe^{2+}$ and $Fe^{3+}$ as they react (respectively) with potassium ferricyanide and potassium ferrocyanide to form Prussian blue.

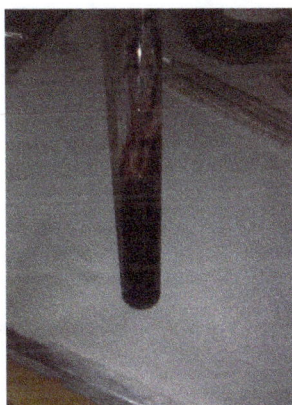

Blood-red positive thiocyanate test for iron (III)

Iron (III) complexes are quite similar to those of chromium (III) with the exception of iron (III)'s preference for *O*-donor instead of *N*-donor ligands. The latter tend to be rather more unstable than iron (II) complexes and often dissociate in water. Many Fe–O complexes show intense colors and are used as tests for phenols or enols. For example, in the ferric chloride test, used to determine the presence of phenols, iron (III) chloride reacts with a phenol to form a deep violet complex:

$$3 \text{ ArOH} + FeCl_3 \rightarrow Fe(OAr)_3 + 3 \text{ HCl (Ar = aryl)}$$

Among the halide and pseudohalide complexes, fluoro complexes of iron (III) are the most stable, with the colorless $[FeF_5(H_2O)]^{2-}$ being the most stable in aqueous solution. Chloro complexes are less stable and favor tetrahedral coordination as in $[FeCl_4]^-$; $[FeBr_4]^-$ and $[FeI_4]^-$ are reduced easily to iron(II). Thiocyanate is a common test for the presence of iron (III) as it forms the blood-red $[Fe(SCN)(H_2O)_5]^{2+}$. Like manganese (II), most iron (III) complexes are high-spin, the exceptions being those with ligands that are high in the spectrochemical series such as cyanide. An example of a low-spin iron (III) complex is $[Fe(CN)_6]^{3-}$. The cyanide ligands may easily be detached in

$[Fe(CN)_6]^{3-}$, and hence this complex is poisonous, unlike the iron(II) complex $[Fe(CN)_6]^{4-}$ found in Prussian blue, which does not release hydrogen cyanide except when dilute acids are added. Iron shows a great variety of electronic spin states, including every possible spin quantum number value for a d-block element from 0 (diamagnetic) to $\frac{5}{2}$ (5 unpaired electrons). This value is always half the number of unpaired electrons. Complexes with zero to two unpaired electrons are considered low-spin and those with four or five are considered high-spin.

Iron (II) complexes are less stable than iron (III) complexes but the preference for $O$-donor ligands is less marked, so that for example $[Fe(NH_3)_6]^{2+}$ is known while $[Fe(NH_3)_6]^{3+}$ is not. They have a tendency to be oxidized to iron (III) but this can be moderated by low pH and the specific ligands used.

## Organometallic Compounds

Iron pentacarbonyl

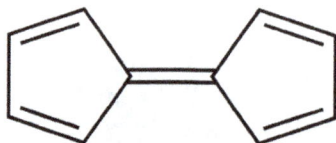

Fulvalene, which Pauson and Kealy sought to prepare

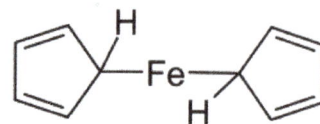

The (incorrect) structure for ferro-cene that Pauson and Kealy proposed

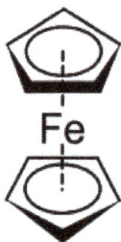

The structural formula of ferrocene

Powdered ferrocene

Cyanide complexes are technically organometallic but more important are carbonyl complexes and sandwich and half-sandwich compounds. The premier iron(0) compound is iron pentacarbonyl, $Fe(CO)_5$, which is used to produce carbonyl iron powder, a highly reactive form of metallic iron. Thermolysis of iron pentacarbonyl gives the trinuclear cluster, triiron dodecacarbonyl. Collman's reagent, disodium tetracarbonylferrate, is a useful reagent for organic chemistry; it contains iron in the −2 oxidation state. Cyclopentadienyliron dicarbonyl dimer contains iron in the rare +1 oxidation state.

Ferrocene was an extremely important compound in the early history of the branch of organometallic chemistry, and to this day iron is still one of the most important metals in this field. It was first synthesised in 1951 during an attempt to prepare the fulvalene ($C_{10}H_8$) by oxidative dimerization of cyclopentadiene; the resultant product was found to have molecular formula $C_{10}H_{10}Fe$ and reported to exhibit "remarkable stability". The discovery sparked substantial interest in the field of organometallic chemistry, in part because the structure proposed by Pauson and Kealy (shown at right) was inconsistent with then-existing bonding models and did not explain its unexpected stability. Consequently, the initial challenge was to definitively determine the structure of ferrocene in the hope that its bonding and properties would then be understood. The shockingly novel sandwich structure, $[Fe(\eta^5\text{-}C_5H_5)_2]$, was deduced and reported independently by three groups in 1952: Robert Burns Woodward and Geoffrey Wilkinson investigated the reactivity in order to determine the structure and demonstrated that ferrocene undergoes similar reactions to a typical aromatic molecule (such as benzene), Ernst Otto Fischer deduced the sandwich structure and also began synthesising other metallocenes including cobaltocene; Eiland and Pepinsky provided X-ray crystallographic confirmation of the sandwich structure.

Applying valence bond theory to ferrocene by considering an $Fe^{2+}$ centre and two cyclopentadienide anions $(C_5H_5^-)$, which are known to be aromatic according to Hückel's rule and hence highly stable, allowed correct prediction of the geometry of the molecule. Once molecular orbital theory was successfully applied and the Dewar-Chatt-Duncanson model proposed, the reasons for ferrocene's remarkable stability became clear. Ferrocene was not the first organometallic compound known – Zeise's salt, $K[PtCl_3(C_2H_4)]\cdot H_2O$ was reported in 1831 and Mond's discovery of $Ni(CO)_4$ occurred in 1888, but it was ferrocene's discovery that began organometallic chemistry as a separate area of chemistry. It was so important that Wilkinson and Fischer shared the 1973 Nobel Prize for Chemistry "for their pioneering work, performed independently, on the chemistry of the organometallic, so called sandwich compounds". Ferrocene itself can be used as the backbone of a ligand, e.g. 1,1'-bis(diphenylphosphino)ferrocene (dppf). Ferrocene can itself be oxidized to the ferrocenium cation $(Fc^+)$; the ferrocene/ferrocenium couple is often used as a reference in electrochemistry.

Metallocenes like ferrocene can be prepared by reaction of freshly-cracked cyclopentadiene with iron(II) chloride and base. It is an aromatic substance and undergoes substitution reactions rather than addition reactions on the cyclopentadienyl ligands. For example, Friedel-Crafts acylation of ferrocene with acetic anhydride yields acetylferrocene just as acylation of benzene yields acetophenone under similar conditions.

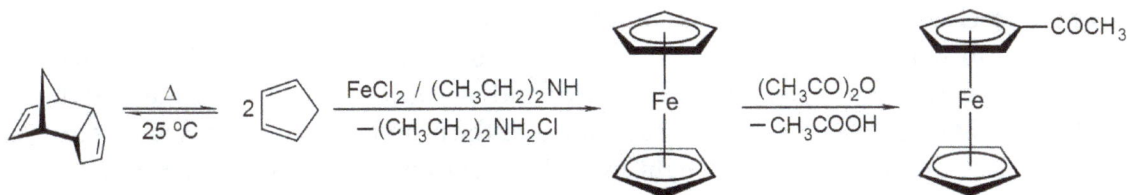

Iron-centered organometallic species are used as catalysts. The Knölker complex, for example, is a transfer hydrogenation catalyst for ketones.

## Wrought Iron

Wrought iron is an iron alloy with very low carbon content with respect to cast iron. It is soft, ductile, magnetic, and has high elasticity and tensile strength. It can be heated and reheated and worked into various shapes.

Although wrought iron exhibits properties that are not found in other forms of ferrous metal, it lacks the carbon content necessary for hardening through heat treatment. Wrought iron may be welded in the same manner as mild steel, but the presence of oxides or inclusions will provide defective results.

## Preparation of Wrought Iron

It is prepared from pig iron by burning out C, Si, Mn, P and sulphur in a puddling furnace. So wrought iron is a purer form of pig iron. Pig iron contains 6% or more of these impurities but their percentage is reduced to about one per cent in wrought iron. Carbon content is reduced to about 0.02%.

In the process of purification of pig iron into wrought iron, a minute quantity of slag is incorporated into wrought iron and is uniformly distributed in it. The presence of slag gives fibrous structure to wrought iron.

## Properties

The microstructure of wrought iron, showing dark slag inclusions in ferrite

The slag inclusions, or stringers, in wrought iron give it properties not found in other forms of ferrous metal. There are approximately 250,000 inclusions per square inch. A fresh fracture shows a clear bluish color with a high silky luster and fibrous appearance.

Wrought iron lacks the carbon content necessary for hardening through heat treatment, but in areas where steel was uncommon or unknown, tools were sometimes cold-worked (hence cold iron) in order to harden them. An advantage of its low carbon content is its excellent weldability. Furthermore, sheet wrought iron cannot bend as much as steel sheet metal (when cold worked). Wrought iron can be melted and cast, however the product is no longer wrought iron, since the slag stringers characteristic of wrought iron disappear on melting, so the product resembles impure cast Bessemer steel. There is no engineering advantage as compared to cast iron or steel, both of which are cheaper.

Due to the variations in iron ore origin and iron manufacture, wrought iron can be inferior or superior in corrosion resistance compared to other iron alloys. There are many mechanisms behind that corrosion resistance. Chilton and Evans found that nickel enrichment bands reduce corrosion. They also found that in puddled, forged and piled iron, the working-over of the metal spread out copper, nickel and tin impurities, which produces electrochemical conditions that slow down corrosion. The slag inclusions have been shown to disperse corrosion to an even film, enabling the iron to resist pitting. Another study has shown that slag inclusions are pathways to corrosion. Other studies show that sulfur impurities in the wrought iron decrease corrosion resistance, but phosphorus increase corrosion resistance. Environments with a high concentration of chlorine ions also decreases wrought iron's corrosion resistance.

Wrought iron may be welded in the same manner as mild steel, but the presence of oxide or inclusions will give defective results. The material has a rough surface, so it can hold platings and coatings better. For instance, a galvanic zinc finish applied to wrought iron is approximately 25–40% thicker than the same finish on steel. In table, the chemical composition of wrought iron is compared to that of pig iron and carbon steel. Although it appears that wrought iron and plain carbon steel have similar chemical compositions that is deceiving. Most of the manganese, sulfur, phosphorus, and silicon are incorporated into the slag fibers present in the wrought iron, so, really, wrought iron is purer than plain carbon steel.

| Table: Chemical composition comparison of pig iron, plain carbon steel, and wrought iron | | | | | | |
|---|---|---|---|---|---|---|
| Material | Iron | Carbon | Manganese | Sulfur | Phosphorus | Silicon |
| Pig iron | 91–94 | 3.5–4.5 | 0.5–2.5 | 0.018–0.1 | 0.03–0.1 | 0.25–3.5 |
| Carbon steel | 98.1–99.5 | 0.07–1.3 | 0.3–1.0 | 0.02–0.06 | 0.002–0.1 | 0.005–0.5 |
| Wrought iron | 99–99.8 | 0.05–0.25 | 0.01–0.1 | 0.02–0.1 | 0.05–0.2 | 0.02–0.2 |
| All units are percent weight | | | | | | |

| Table: Properties of wrought iron | |
|---|---|
| Property | Value |
| Ultimate tensile strength [psi (MPa)] | 34,000–54,000 (234–372) |
| Ultimate compression strength [psi (MPa)] | 34,000–54,000 (234–372) |
| Ultimate shear strength [psi (MPa)] | 28,000–45,000 (193–310) |
| Yield point [psi (MPa)] | 23,000–32,000 (159–221) |
| Modulus of elasticity (in tension) [psi (MPa)] | 28,000,000 (193,100) |
| Melting point [°F (°C)] | 2,800 (1,540) |
| Specific gravity | 7.6–7.9 |
| | 7.5–7.8 |

Amongst its other properties, wrought iron becomes soft at red heat, and can be easily forged and forge welded. It can be used to form temporary magnets, but cannot be magnetized permanently, and is ductile, malleable and tough.

## Ductility

For most purposes, ductility is a more important measure of the quality of wrought iron than tensile strength. In tensile testing, the best irons are able to undergo considerable elongation before failure. Higher tensile wrought iron is brittle.

## Applications

Wrought iron furniture has a long history, dating back to Roman times. There are 13th-century wrought iron gates in Westminster Abbey in London, and wrought iron furniture appeared to reach its peak popularity (in Britain) in the 17th century during the reign of William III and Mary II. However, cast iron and cheaper steel caused a gradual decline in wrought iron manufacture; the last wrought ironworks in Britain closed in 1974.

It is also used to make home decor items such as baker's racks, wine racks, pot racks, etageres, table bases, desks, gates, beds, candle holders, curtain rods, bars and bar stools.

The vast majority of wrought iron available today is from reclaimed materials. Old bridges and anchor chains dredged from harbors are major sources. The greater corrosion resistance of wrought iron is due to the siliceous impurities (naturally occurring in iron ore), namely ferric silicate.

The use of wrought iron today is usually reserved for special applications, such as fine carpentry tools and historical restoration for objects of great importance.

## Cast Iron

Cast iron is obtained from the pig-iron which is re-melted with coke and limestone. Pig iron is nothing but impure iron which is obtained from the iron ore. Cast iron has lot of engineering properties so, that it can be used in many ways like for sanitary fittings, rail chairs, casting molds etc.

## Manufacturing of Cast Iron

Cast iron in manufactured from re-melting process as mentioned above. This process takes place in a furnace called couple furnace. The furnace is 5 meters in height and cylindrical in shape with 1 m diameter.

The raw materials pig iron, coke and lime stone are entered from the charging door of chamber which provided at the top. Air blast is introduced into the chamber using air blast inlet which removes the impurities present in the pig iron.

Hence, pure cast iron is obtained from the bottom out let and it is poured into molds of required shape. These molds are called as cast iron castings.

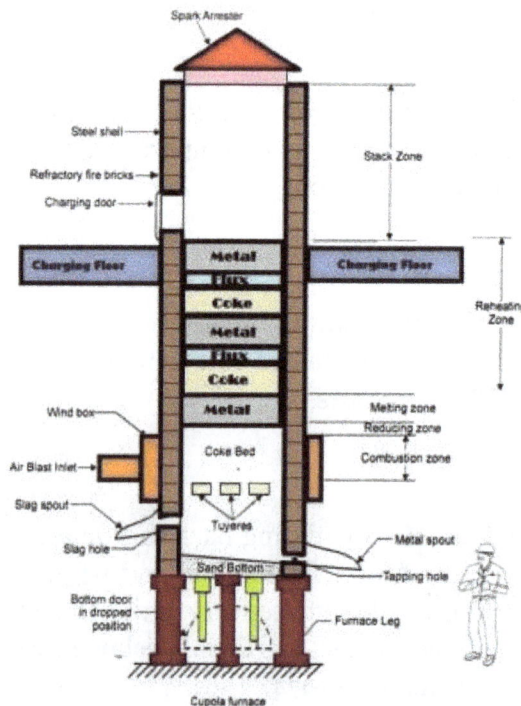

Cupola furnace

## Types of Cast Iron used as a Building Material in Construction Works

Following are the types of cast iron used as a building material in construction and their uses:

- Grey cast iron

- Malleable cast iron

- Mottled cast iron

- Toughened cast iron

- White cast iron

- Ductile cast iron

- Chilled cast iron

### Grey Cast Iron

As the name suggests, it is grey in color. It has coarse crystalline structure. Its melting point is very low thus it has weak strength and it is only used for casting purposes.

### Malleable Cast Iron

Malleability is the property which helps the materials to undergo any shape without breaking or cracking. Hence, malleable cast iron is used for making many types of materials. It has good corrosive resistance.

Its manufacturing process involves two steps. In the first step, it is casted and cooled as ordinary cast iron and then again it is heated to 1050°C and soaked in water for long period (several hours or days).

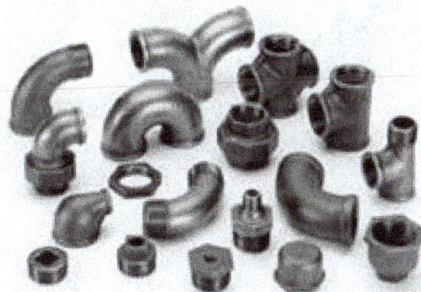

Malleable Cast Iron fittings black range

Hence, carbon content is slightly reduced and graphite content is precipitated as temper carbon. This reduces the brittleness of cast iron. So, it can be worked easily using machines. It is used for making pipe fittings, fastenings, automobiles etc.

## Mottled Cast Iron

Mottled iron is the medium stage cast iron which properties are in between grey cast iron and white cast iron. It has Small amount of graphite in its composition. So, mottled type fractures are developed in its micro structure.

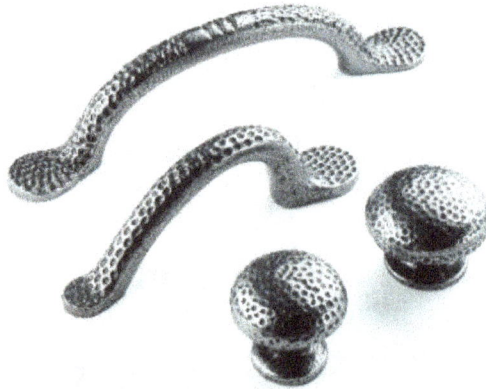

## Toughened Cast Iron

Toughened cast iron is the combination of cast iron and wrought iron. To obtain this wrought iron scrap and cast iron melted together. The composition of wrought iron is about 0.15 to 0.25 weight of the cast iron.

## White Cast Iron

It is in silver color. Its melting point is high so, strength wise it is better but not used for delicate casting purposes. Because of its heavy strength, it cannot be used easily.

## Ductile Cast Iron

Ductile cast iron is also called as spheroidal graphite iron. Its manufacturing process is very easy compared to other types.

Its manufacturing process consists manganese treatment which helps to increase the carbon content and opposes the formation of graphite in flaky form. It has very good engineering properties than malleable cast iron.

Ductile cast iron has very good corrosion resistance, high strength and durability. So, usage of ductile iron dominates the other types. It is used for making sewer pipes, water conveying pipes etc.

## Chilled Cast Iron

Chilled cast iron consists two layers of which one layer has white cast iron properties and other one has grey cast iron properties. This type of iron is used for casting process in which grey cast iron layer is provided in inner surface and white cast iron layer is provided as outer surface. Hence the casting molds serve longer. Machine parts are also made using chilled cast iron.

## Composition of Different Types of Cast Iron

Different types of cast iron are composed of different quantities of constituents shown in table below:

| Cast Iron type | Percentage of constituents (%) | | | | |
|---|---|---|---|---|---|
| | Carbon | Silica | Manganese | Sulphur | Phosphorus |
| Grey cast iron | 2.5-4.0 | 1-3 | 0.2-1 | 0.02-0.25 | 0.02-1 |
| White cast iron | 1.8-3.6 | 0.5-1.9 | 0.25-0.8 | 0.06-0.2 | 0.06-0.2 |
| Ductile cast iron | 3-4 | 1.8-2.8 | 0.1-1.0 | 0.01-0.03 | 0.01-0.1 |
| Malleable cast iron | 2.0-2.9 | 0.9-1.9 | 0.15-1.2 | 0.02-0.2 | 0.02-0.2 |
| Mottled cast iron | 2.5-4.0 | 1.0-3.0 | 0.2-1.0 | 0.01-0.03 | 0.01-0.1 |

## Properties of Cast Iron as a Building Material

The properties of cast iron are as follows:

- Good corrosion resistance, so it can be used for water carrying pipes etc.

- Does not get attracted to magnet.

- Specific gravity is 7.5.

- Melting point is about 1250°C.

- Tensile strength is about 150 N/mm2 and compressive strength is about 600 N/mm2.so, it is good in compression.

- It becomes soft when placed in salt water and it shrinks on cooling.

- It cannot be useful for forging work because of lack of plasticity.

## Uses of Cast Iron in Building Construction

Cast iron can be used for making different tools, materials etc. as described below:

- Many types of sanitary fittings like manholes, sewer pipes, water pipes, cisterns are manufactured using cast iron.

- Metal columns and column bases can be made using cast iron.

- Casting molds for making metal staircases, lamp posts, gates etc. are made using cast iron.

- Rail chairs and carriage wheels are manufactured.

- Several types of agricultural implements can be made.

- Machinery parts can be manufactured but shock cannot be resisted by cast iron.

## Structural Steel

Structural steel is a category of steel construction mate that is produced with a particular cross section or shape, and some specified values of strength and chemical composition. Structural steel composition, strength, size, shape, strength, and storage are controlled in most advanced countries. The word structural steel includes a broad variety of low carbon and manganese steels that are used in great numbers for civil and marine engineering applications. Numerous structural steels also include minor quantities of significant additions of other elements like Nb, V, Ti and Al. These are called High Strength Low Alloy or micro-alloyed steels. Structural steels are manufactured in section and plate shapes and are normally used in bridges, buildings, ships, and pipelines.

## Types of Structural Steel

After iron, carbon is the most important element in steel. The increase of carbon produces materials with high strength and low ductility. The techniques used for the production of steel are high- computerized stress analysis, precision stress analysis, and innovative jointing. The types of structural steel sections normally used are beams, channels, flats, and angles. The main kinds of structural steel are generally categorized according to the under mentioned categories of chemical composition:

- Carbon-manganese steels: The major chemical ingredients are iron, carbon, and manganese. These are normally called mild structural steels or carbon steels. The strength and ductility are high, and being economical is therefore widely used. The famous category amongst this type is ASTM grade A36.

- High-strength, low-alloy steels: This is a recent development in the steel industry. Chemical elements are added to improve the strength. A commonly used type is ASTM grade A572.

- High-strength tempered and quenched alloy steels: These are used for structural purposes and generally available is ASTM grade A514.

## Fireproofing of Structural Steel

Fire-resistance rating is determined by the time taken for the steel that is being tested to attain the temperature fixed by the standard. Structural steel needs external insulation that is also called fireproofing, to prevent the deterioration of steel in the occurrence of a fire. On heating, the steel expands and becomes softer, and finally the structural integrity is lost. If sufficient energy is provided, steel may also melt. The transfer rate of heat to steel can be reduced by using fireproofing materials. While concrete structures may be able to resist fire damage without extra fireproofing, concrete may deteriorate, particularly if the moisture content is high. Fireproofing is typically used in tunnels and locations where hydrocarbons fire is likely to break out. Fireproofing is incorporated to meet the fire protection necessities that are essential due to the building codes.

While the shorthand version of how structural steel is created involves heating iron up and adding certain substance to achieve specific properties, the long version is much more involved.

Raw iron is the chief ingredient, but it is rarely found pure in nature. Most often it already contains carbon, but usually in too high a concentration. Some carbon needs to be removed, but not all. Because of that, the manufacturing of steel products can be an involved process.

1. First, the raw iron ore is crushed and sorted. There are a number of different refining processes, all designed to sort out the best grades of iron, usually around 60 percent.

2. Ore is the loaded into a blast furnace from the top and heated, while hot air is blown into the furnace from the bottom. The reaction that takes place begins to remove impurities as pure iron sinks to the bottom of the furnace.

3. The molten iron is drawn off and is further heated to allow the inclusion of other substances, such as manganese, that deliver different properties to the finished steel product.

Once the steel has been created, it is formed into a number of different configurations, depending on how it will be used. Beam, channel, angle, plate and hollow steel tube are the most common.

## Fabrication

Even after it has been formed into shape, structural steel still requires fabricating and welding. A skilled fabricator or welder can take advantage of the relative malleability of structural steel to create whatever shape is needed for a specific application.

Welding is largely replacing riveting as the chosen method of fabricating structural steel, and with good reason. Welded structures are:

1.  Lighter than similarly riveted structures

2.  More cost effective

3.  Easier to navigate and mold than their riveted counterparts

Design and construction of steel structures depends on the properties of structural structures. Different properties of steel and their importance in design and construction of steel structures are discussed.

Fig: Structural Steel Design and Construction

## Properties of Structural Steel for Design and Construction of Steel Structures

Properties of structural steel include:

* Tensile properties

* Shear properties

* Hardness

* Creep

* Relaxation

* Fatigue

## Tensile Properties of Structural Steel

There are different categories of steel structures which can be used in the construction of steel buildings. Typical stress strain curves for various classes of structural steel, which are derived from steel tensile test, are shown in figure.

The initial part of the curve represents steel elastic limit. In this range, steel structure deformation is not permanent, and the steel regain its original shape upon the removal of the load.

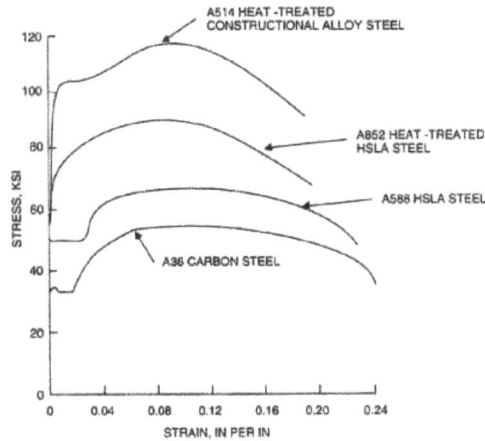

Figure: Typical Stress Strain Curve Different Classes of Structural Steel

The elastic modulus of all steel classes is same and equal to 200000MPa or $2 \times 10^6$ MPa. As the load on the steel is increased, it would yield at a certain point after which plastic range will be reached.

The yield point is the point at which steel specimen reach 0.002 strain under the effect of specific stress (yield stress).

Ductility of steel structure as shown in figure is crucial properties that allow redistribution of stress in continuous steel elements. Ductility is expressed by percentage of steel cross sectional reduction.

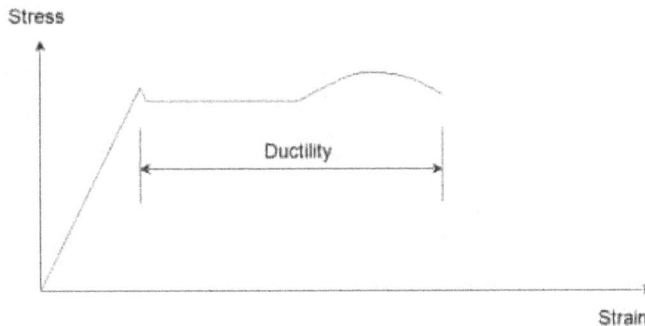

Figue: Stress Strain Curve of Structural Steel

As far as poisons ratio is concerned, it is the ratio of transverse strain to axial strain and it is about 0.30 and 0.50 in elastic and plastic range, respectively.

Regarding cold working of structural steel, it is the process in which different shapes of steel structure are produced at room temperature.

Consequently, steel structure ductility is increased but its ductility is reduced. Residual stress is a stress that stays in steel element after it has been fabricated.

It is necessary to consider strain rate while tensile test is conducted because it modifies steel tensile properties.

If steel structure is used for dynamic loads, then high strain rate would be considered. However, normal strain rate is adopted for steel used in the construction of structure designed for static loads.

The ability of steel structure to accommodate energy is called steel toughness.

## Shear Properties of Structural Steel

Shear strength of steel structure is specified at the failure under shear stress and it is about 0.57 times yield stress of structural steel.

Regarding elastic shear modulus, it is expressed as the ratio of shear stress to shear strain in elastic range of steel structure.

Commonly, elastic shear modulus of steel structure can be taken as 75.84Gpa or the following formula can be used to compute elastic shear modulus.

$$G = \frac{E}{2(1+\mu)}$$

Where:

$G$: Steel structure shear elastic modulus

$E$: Modulus of elasticity of steel structure

$\mu$: Position's ratio

## Hardness of Structural Steel

Hardness is the measure of ability of steel structure to withstand inelastic deformation. Standard test methods and definitions for mechanical testing of steel products (A370-05) specify three different tests to evaluate steel hardness namely: Brinell, Rockwell and portable.

Any of these tests can be used to estimate steel structure hardness. Not only is the steel structure hardness used to examine the uniformity of different products but also to evaluate steel tensile strength.

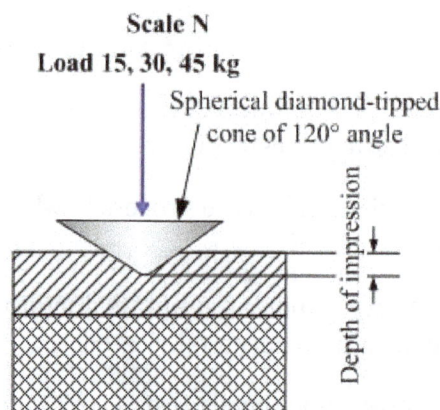

**Scale N**
**Load 15, 30, 45 kg**

Spherical diamond-tipped cone of 120° angle

Depth of impression

Figure: Rockwell Test for Structural Steel Hardness Evaluation

## Creep of Structural Steel Relaxation

Creep is gradual variation of strain of steel structure under constant stress. It occurs due to the influence of constant stress and the effect of fire.

Creep property is insignificant for structural steel frame design and construction apart from the case in which the effect of fire should be taken into consideration.

## Structural Steel Relaxation

It is a step by step reduction of structural steel under a constant stress. Usually, yield strength of steel structure increases around 5% over stress relieved strain and the steel structure would suffer from plastic elongation which around 0.01.

## Fatigue of Structural Steel

Fatigue is the failure of steel structure due to crack initiation and development under the influence of cyclic loading. Various tests are available to evaluate structural steel fatigue such as flexure test, rotating beam test and axial load test.

Figure: Fatigue Test of Structural Steel

## Concrete

Concrete is a mixture of paste and aggregates. The paste, composed of portland cement and water, coats the surface of the fine and coarse aggregates. Through a chemical reaction called hydration, the paste hardens and gains strength to form the rock-like mass known as concrete.

Within this process lies the key to a remarkable trait of concrete: it's plastic and malleable when newly mixed, strong and durable when hardened.

Concrete's durability, strength and relatively low cost make it the backbone of buildings and infrastructure worldwide—houses, schools and hospitals as well as airports, bridges, highways and rail systems. The most-produced material on Earth will only be more in demand as, for example, developing nations become increasingly urban, extreme weather events necessitate more durable building materials and the price of other infrastructure materials continues to rise.

Even construction professionals sometimes incorrectly use the terms cement and concrete interchangeably. Cement is actually an ingredient of concrete. It is the fine powder that, when mixed with water, sand, and gravel or crushed stone (fine and coarse aggregate), forms the rock-like mass known as concrete.

## The Forms ofs Concrete

Concrete is produced in four basic forms, each with unique applications and properties.

1.  Ready-mixed concrete, far the most common form, accounts for nearly three-fourths of all concrete. It's batched at local plants for delivery in the familiar trucks with revolving drums.

2.  Precast concrete products are cast in a factory setting. These products benefit from tight quality control achievable at a production plant. Precast products range from concrete bricks and paving stones to bridge girders, structural components, and wall panels. Concrete masonry another type of manufactured concrete, may be best known for its conventional 8-by-8-by-16-inch block. Today's masonry units can be molded into a wealth of shapes, configurations, colors, and textures to serve an infinite spectrum of building applications and architectural needs.

3.  Cement-based materials represent products that defy the label of "concrete," yet share many of its qualities. Conventional materials in this category include mortar, grout, and terrazzo. soil-cement and roller-compacted concrete —"cousins" of concrete—are used for pavements and dams. Other products in this category include flow able fill and cement-treated bases.

4.  A new generation of advanced products incorporates fibers and special aggregate to create roofing tiles, shake shingles, lap siding, and countertops.

## Grade of Concrete

Grade of concrete denotes its strength required for construction. For example, M30 grade signifies that compressive strength required for construction is 30MPa. The first letter in grade "M" is the mix and 30 is the required strength in MPa.

Based on various lab tests, grade of concrete is presented in mix proportions. For example, for M30 grade, the mix proportion can be 1:1:2, where 1 is the ratio of cement, 1 is the ratio of sand and 2 is the ratio of coarse aggregate based on volume or weight of materials.

The strength is measured with concrete cube or cylinders by civil engineers at construction site. Cube or cylinders are made during casting of structural member and after hardening it is cured for 28 days. Then compressive strength test is conducted to find the strength.

Regular grades of concrete are M15, M20, M25 etc. For plain cement concrete works, generally M15 is used. For reinforced concrete construction minimum M20 grade of concrete are used.

| Concrete Grade | Mix Ratio | Compressive Strength | |
|---|---|---|---|
| | | MPa (N/mm²) | psi |
| **Normal Grade of Concrete** | | | |
| M5 | 1 : 5 : 10 | 5 MPa | 725 psi |
| M7.5 | 1 : 4 : 8 | 7.5 MPa | 1087 psi |
| M10 | 1 : 3 : 6 | 10 MPa | 1450 psi |
| M15 | 1 : 2 : 4 | 15 MPa | 2175 psi |
| M20 | 1 : 1.5 : 3 | 20 MPa | 2900 psi |
| **Standard Grade of Concrete** | | | |
| M25 | 1 : 1 : 2 | 25 MPa | 3625 psi |
| M30 | Design Mix | 30 MPa | 4350 psi |
| M35 | Design Mix | 35 MPa | 5075 psi |
| M40 | Design Mix | 40 MPa | 5800 psi |
| M45 | Design Mix | 45 MPa | 6525 psi |
| **High Strength Concrete Grades** | | | |
| M50 | Design Mix | 50 MPa | 7250 psi |
| M55 | Design Mix | 55 MPa | 7975 psi |
| M60 | Design Mix | 60 MPa | 8700 psi |
| M65 | Design Mix | 65 MPa | 9425 psi |
| M70 | Design Mix | 70 MPa | 10150 psi |

## Reinforced Concrete

Reinforced concrete, or RCC, is concrete that contains embedded steel bars, plates, or fibers that strengthen the material. The capability to carry loads by these materials is magnified, and because of this RCC is used extensively in all construction. In fact, it has become the most commonly utilized construction material.

Reinforced materials are embedded in the concrete in such a way that the two materials resist the applied forces together. The compressive strength of concrete and the tensile strength of steel form a strong bond to resist these stresses over a long span. Plain concrete is not suitable for most construction projects because it cannot easily withstand the stresses created by vibrations, wind, or other forces.

## Performance of Reinforced Concrete

Concrete consists of a cement and stone aggregate mixture that forms a rigid structure with the addition of water. When steel that has a high tensile strength is embedded in concrete, the composite material withstands compression, bending, and tensile stresses. Such a material can be used for making any size and shape, for utilization in the construction.

The main quality of reinforced concrete is similarity of its coefficient of thermal expansion with that of steel, due to which the internal stresses initiated due to variation in thermal expansion or contraction are eliminated.

Secondly, on the hardening of the cement paste inside the concrete, it corresponds to the surface features of the steel, allowing the stresses to be efficiently transmitted between the two materials.

The cohesive characteristics between the steel and concrete are enhanced by the roughening of steel bars.

Thirdly, a film is formed on the steel because of the alkaline environment created by lime, due to which the steel becomes extra-resistant to corrosion.

Reinforced concrete is used in almost all construction projects. It has also surpassed wood to become the most commonly used construction material.

## Steel Reinforced Concrete

All kinds of materials can be used to reinforce concrete. But steel is the most popular choice due to its ability to expand and contract at the same rate as the concrete. This feature ensures the steel will not crack the concrete during fluctuations in temperature.

Reinforcing steel is usually embedded in the concrete in the form of steel rods, steel bars, or steel meshes. The steel is tied together to make a skeletal structure before the concrete is poured over top.

After the concrete sets, the newly created reinforced concrete will have superior resisting properties. It can withstand high tensile and compressive stresses for long periods.

## Fiber Reinforced Concrete

Fibers are a new type of reinforcement used to make reinforced concrete. These fibers are rated by their aspect ratio, which is a ratio of the fibers' lengths to its diameter. They are randomly dispersed throughout the concrete and stirred well to create a uniform mix.

The most common material used to make the fibers include glass, synthetic, and even steel. Other materials used to make the fibers that are not so common include asbestos, which is economical, and carbon, which has great mechanical properties.

## Properties of Reinforced Concrete

Concrete by itself is an aggregate mixture of cement and stone. This mixture does not stick together when dry but forms a rigid structure when water is added and then stirred.

With the addition of steel inside the concrete, the newly formed reinforced concrete has a coefficient of thermal expansion that is similar with that of both the steel and the concrete. As a result, internal stresses resulting from variations in the environmental temperature are almost non-existent.

Additionally, when the concrete cement hardens, it corresponds to the surface features of the steel. This allows the stresses acting around the reinforced concrete to be efficiently spread between the two materials.

Another desirable property of reinforced concrete is the development of a thin film on the surface of the steel. This occurs due to the alkaline environment caused by lime. As a result, the steel becomes extra-resistant to corrosion as moisture cannot penetrate this layer of lime easily.

## Problems with Reinforced Concrete

The most common problem associated with reinforced concrete is corrosion. In the presence of moisture, the embedded steel may corrode and chip away. Corrosion results in extensive damage to the inner core of the concrete. For large structures such as high-rises, bridges, and dams, this presents a serious problem, as the failure of such a large structure would be catastrophic.

Techniques have been developed to detect the extent of the corrosion in the embedded steel. This information can be used to calculate the lifespan of large reinforced concrete structures so they can be rebuilt at the end of their life cycle.

## Prestressed Concrete

Prestressed concrete is a structural material that allows for predetermined, engineering stresses to be placed in members to counteract the stresses that occur when they are subject to loading. It combines the high strength compressive properties of concrete with the high tensile strength of steel.

In ordinary reinforced concrete, stresses are carried by the steel reinforcement, whereas prestressed concrete supports the load by induced stresses throughout the entire structural element. This makes it more resistant to shock and vibration than ordinary concrete, and able to form long, thin structures with much smaller sectional areas to support equivalent loads.

## Steel

Steel used for prestressing may be in the form of wire or tendons that can be grouped to form cables. Solid bars may also be used.

Wire is made by cold-drawing a high carbon steel rod through a series of reducing dies. The wire diameter typically ranges from 3-7 mm and may be round, crimped or indented to give it better bond strength. Another form of tendon is strand which consists of a straight core wire around which is wound in helixes around further wires to give formats such as 7 wire (6 over 1) and 19 wire (9 over 9 over 1). Similar to wire tendons, strand can be used individually or in groups to form cables.

## Method

The process of prestressed concrete can be either through pre-tensioning or post-tensioning.

## Pre-tensioning

This process involves the stressing of wires or cables by anchoring them at the end of a metal form, which may be up to 120 m in length. Hydraulic jacks stress the wire as required, often adding 10% to accommodate creep and other pre-stress losses that may be incurred. Side moulds are then fixed and the concrete placed around the tensioned wires. The concrete hardens and shrinks, gripping the steel along its length, transferring the tension from the jacks to exert a compressive force in the concrete.

Once the concrete has reached the desired strength, the tensioned wires are released from the jacks. A typical concrete strength of 28 N/mm2 can be achieved by 24-hour steam curing, as well as using additives.

To create shorter members, dividing plates can be placed at any point along the member which, when removed, permit the cutting of the wires.

## Post-tensioning

This follows the reverse method to pre-tensioning, whereby the concrete member is cast and the prestressing occurs after the concrete is hardened. This method is often used where stressing is to be carried out on site after casting an insitu component or where a series of precast concrete units are to be joined together to form the required member.

The wires, cables or bars may be positioned in the unit before concreting, but bonding to the concrete is prevented by using a flexible duct or rubber sheath which is deflated and removed when the concrete has hardened.

Stressing is carried out after the concrete has been cured by means of hydraulic jacks operating from one or both ends of the member. Due to the high local stresses at the anchorage positions it is common for a helical (spiral) reinforcement to be included in the design. When the required stress has been reached, the wire or cables are anchored to maintain the prestress. The ends of the unit are sealed with cement mortar to prevent corrosion due to any entrapped moisture and to assist in stress distribution.

Anchorages used in post-tensioning depend on whether the tendons are to be stressed individually or as a group. Most systems use a form of split cone wedges or jaws which act against a form of bearing or pressure plate.

There are many different post-tensioning systems. For example, the Freyssinet system enables the stressing strands to be tensioned simultaneously using centre hole tensioning jacks, anchored by tapered jaws. This is suitable for pre-stressing elements up to 50 m in length.

The Macalloy system on the other hand, involves applying stress to the concrete by means of a solid bar, usually with a diameter of 25-75 mm. The bar is anchored at each end by a special nut which bears against an end plate to distribute the load.

## Advantages and Disadvantages

The advantages of prestressed concrete include:

- The inherent compressive strength of concrete is used to its fullest.

- The special alloy steels used to form the prestressing tendons are used to their fullest.

- Tension cracks are eliminated, reducing the risk of the steel components corroding.

- Shear stresses are reduced.

- For any given span and loading condition a reduction in weight can be achieved from using a component with a smaller cross section.

- A composite member can be formed by joining individual precast concrete units together.

The disadvantages of prestressed concrete include:

- A high degree of workmanship and control is required.

- Special alloy steels are more expensive than traditional steels used in reinforced concrete.

- Expensive equipment is needed and there are complex safety requirements.

## Aluminium

Aluminium is a lightweight, silvery-white metal of main Group 13 (IIIa, or boron group) of the periodic table. Aluminium is the most abundant metallic element in Earth's crust and the most widely used nonferrous metal. Because of its chemical activity, aluminium never occurs in the metallic form in nature, but its compounds are present to a greater or lesser extent in almost all rocks, vegetation, and animals. Aluminium is concentrated in the outer 10 miles (16 km) of Earth's crust, of which it constitutes about 8 percent by weight; it is exceeded in amount only by oxygen and silicon. The name aluminium is derived from the Latin word alumen, used to describe potash alum, or aluminium potassium sulfate, $KAl(SO_4)_2 \cdot 12H_2O$.

| Element Properties | |
|---|---|
| atomic number | 13 |
| atomic weight | 26.9815 |
| melting point | 660 °C (1,220 °F) |
| boiling point | 2,467 °C (4,473 °F) |
| specific gravity | 2.70 (at 20 °C [68 °F]) |
| valence | 3 |
| electron configuration | $1s^2 2s^2 2p^6 3s^2 3p^1$ |

## Boron Group Element

boron group element are boron (B), aluminium (Al), gallium (Ga), indium (In), thallium (Tl), and nihonium (Nh). They are characterized as a group by having three electrons in the outermost parts of their atomic structure. Boron, the lightest.

**Periodic table of the elements**

*Numbering system adopted by the International Union of Pure and Applied Chemistry (IUPAC).

## Occurrence, uses, and Properties

Aluminium occurs in igneous rocks chiefly as aluminosilicates in feldspars, feldspathoids, and micas; in the soil derived from them as clay; and upon further weathering as bauxite and iron-rich laterite. Bauxite, a mixture of hydrated aluminium oxides, is the principal aluminium ore. Crystalline aluminium oxide (emery, corundum), which occurs in a few igneous rocks, is mined as a natural abrasive or in its finer varieties as rubies and sapphires. Aluminium is present in other gemstones, such as topaz, garnet, and chrysoberyl. Of the many other aluminium minerals, alunite and cryolite have some commercial importance.

Aluminium is added in small amounts to certain metals to improve their properties for specific uses, as in aluminium bronzes and most magnesium-base alloys; or, for aluminium-base alloys, moderate amounts of other metals and silicon are added to aluminium. The metal and its alloys are used extensively for aircraft construction, building materials, consumer durables (refrigerators, air conditioners, cooking utensils), electrical conductors, and chemical and food-processing equipment.

Pure aluminium (99.996 percent) is quite soft and weak; commercial aluminium (99 to 99.6 percent pure) with small amounts of silicon and iron is hard and strong. Ductile and highly malleable, aluminium can be drawn into wire or rolled into thin foil. The metal is only about one-third as dense as iron or copper. Though chemically active, aluminium is nevertheless highly corrosion-resistant, because in air a hard, tough oxide film forms on its surface.

Aluminium is an excellent conductor of heat and electricity. Its thermal conductivity is about one-half that of copper; its electrical conductivity, about two-thirds. It crystallizes in the face-centred cubic structure. All natural aluminium is the stable isotope aluminum-27. Metallic aluminium and its oxide and hydroxide are nontoxic.

Aluminium is slowly attacked by most dilute acids and rapidly dissolves in concentrated hydrochloric acid. Concentrated nitric acid, however, can be shipped in aluminium tank cars because it renders the metal passive. Even very pure aluminium is vigorously attacked by alkalies such as sodium and potassium hydroxide to yield hydrogen and the aluminate ion. Because of its great affinity for oxygen, finely divided aluminium, if ignited, will burn in carbon monoxide or carbon dioxide with the formation of aluminium oxide and carbide, but, at temperatures up to red heat, aluminium is inert to sulfur.

Aluminium can be detected in concentrations as low as one part per million by means of emission spectroscopy. Aluminium can be quantitatively analysed as the oxide (formula $Al_2O_3$) or as a derivative of the organic nitrogen compound 8-hydroxyquinoline. The derivative has the molecular formula $Al(C_9H_6ON)_3$.

## Physical Properties of Aluminium

### Density of Aluminium

Aluminum has a density around one third that of steel or copper making it one of the lightest commercially available metals. The resultant high strength to weight ratio makes it an important structural material allowing increased payloads or fuel savings for transport industries in particular.

### Strength of Aluminium

Pure aluminium doesn't have a high tensile strength. However, the addition of alloying elements like manganese, silicon, copper and magnesium can increase the strength properties of aluminium and produce an alloy with properties tailored to particular applications.

Aluminium is well suited to cold environments. It has the advantage over steel in that its' tensile strength increases with decreasing temperature while retaining its toughness. Steel on the other hand becomes brittle at low temperatures.

### Corrosion Resistance of Aluminium

When exposed to air, a layer of aluminium oxide forms almost instantaneously on the surface of aluminium. This layer has excellent resistance to corrosion. It is fairly resistant to most acids but less resistant to alkalis.

### Thermal Conductivity of Aluminium

The thermal conductivity of aluminium is about three times greater than that of steel. This makes aluminium an important material for both cooling and heating applications such as heat-exchangers. Combined with it being non-toxic this property means aluminium is used extensively in cooking utensils and kitchenware.

### Electrical Conductivity of Aluminium

Along with copper, aluminium has an electrical conductivity high enough for use as an electrical conductor. Although the conductivity of the commonly used conducting alloy (1350) is only around 62% of annealed copper, it is only one third the weights and can therefore conduct twice as much electricity when compared with copper of the same weight.

## Reflectivity of Aluminium

From UV to infra-red, aluminium is an excellent reflector of radiant energy. Visible light reflectivity of around 80% means it is widely used in light fixtures. The same properties of reflectivity makes aluminium ideal as an insulating material to protect against the sun's rays in summer, while insulating against heat loss in winter.

Table. Properties for aluminium.

| Property | Value |
|---|---|
| Atomic Number | 13 |
| Atomic Weight (g/mol) | 26.98 |
| Valency | 3 |
| Crystal Structure | FCC |
| Melting Point (°C) | 660.2 |
| Boiling Point (°C) | 2480 |
| Mean Specific Heat (0-100°C) (cal/g.°C) | 0.219 |
| Thermal Conductivity (0-100°C) (cal/cms. °C) | 0.57 |
| Co-Efficient of Linear Expansion (0-100°C) (x10$^{-6}$/°C) | 23.5 |
| Electrical Resistivity at 20°C ($\Omega$.cm) | 2.69 |
| Density (g/cm³) | 2.6898 |
| Modulus of Elasticity (GPa) | 68.3 |
| Poissons Ratio | 0.34 |

## Mechanical Properties of Aluminium

Aluminium can be severely deformed without failure. This allows aluminium to be formed by rolling, extruding, drawing, machining and other mechanical processes. It can also be cast to a high tolerance.

Alloying, cold working and heat-treating can all be utilised to tailor the properties of aluminium.

The tensile strength of pure aluminium is around 90 MPa but this can be increased to over 690 MPa for some heat-treatable alloys.

Table. Mechanical properties of selected aluminium alloys.

| Alloy | Temper | Proof Stress 0.20% (MPa) | Tensile Strength (MPa) | Shear Strength (MPa) | Elongation A5 (%) | Elongation A50 (%) | Hardness Brinell HB | Hardness Vickers HV | Fatigue Endur. Limit (MPa) |
|---|---|---|---|---|---|---|---|---|---|
| AA1050A | H2 | 85 | 100 | 60 | 12 | | 30 | 30 | |
| | H4 | 105 | 115 | 70 | 10 | 9 | 35 | 36 | 70 |
| | H6 | 120 | 130 | 80 | 7 | | 39 | | |
| | H8 | 140 | 150 | 85 | 6 | 5 | 43 | 44 | 100 |
| | H9 | 170 | 180 | | | 3 | 48 | 51 | |
| | 0 | 35 | 80 | 50 | 42 | 38 | 21 | 20 | 50 |

| | | | | | | | | |
|---|---|---|---|---|---|---|---|---|
| AA2011 | T3 | 290 | 365 | 220 | 15 | 15 | 95 | 100 | 250 |
| | T4 | 270 | 350 | 210 | 18 | 18 | 90 | 95 | 250 |
| | T6 | 300 | 395 | 235 | 12 | 12 | 110 | 115 | 250 |
| | T8 | 315 | 420 | 250 | 13 | 12 | 115 | 120 | 250 |
| AA3103 | H2 | 115 | 135 | 80 | 11 | 11 | 40 | 40 | |
| | H4 | 140 | 155 | 90 | 9 | 9 | 45 | 46 | 130 |
| | H6 | 160 | 175 | 100 | 8 | 6 | 50 | 50 | |
| | H8 | 180 | 200 | 110 | 6 | 6 | 55 | 55 | 150 |
| | H9 | 210 | 240 | 125 | 4 | 3 | 65 | 70 | |
| | 0 | 45 | 105 | 70 | 29 | 25 | 29 | 29 | 100 |
| AA5083 | H2 | 240 | 330 | 185 | 17 | 16 | 90 | 95 | 280 |
| | H4 | 275 | 360 | 200 | 16 | 14 | 100 | 105 | 280 |
| | H6 | 305 | 380 | 210 | 10 | 9 | 105 | 110 | |
| | H8 | 335 | 400 | 220 | 9 | 8 | 110 | 115 | |
| | H9 | 370 | 420 | 230 | 5 | 5 | 115 | 120 | |
| | 0 | 145 | 300 | 175 | 23 | 22 | 70 | 75 | 250 |
| AA5251 | H2 | 165 | 210 | 125 | 14 | 14 | 60 | 65 | |
| | H4 | 190 | 230 | 135 | 13 | 12 | 65 | 70 | 230 |
| | H6 | 215 | 255 | 145 | 9 | 8 | 70 | 75 | |
| | H8 | 240 | 280 | 155 | 8 | 7 | 80 | 80 | 250 |
| | H9 | 270 | 310 | 165 | 5 | 4 | 90 | 90 | |
| | 0 | 80 | 180 | 115 | 26 | 25 | 45 | 46 | 200 |
| AA5754 | H2 | 185 | 245 | 150 | 15 | 14 | 70 | 75 | |
| | H4 | 215 | 270 | 160 | 14 | 12 | 75 | 80 | 250 |
| | H6 | 245 | 290 | 170 | 10 | 9 | 80 | 85 | |
| | H8 | 270 | 315 | 180 | 9 | 8 | 90 | 90 | 280 |
| | H9 | 300 | 340 | 190 | 5 | 4 | 95 | 100 | |
| | 0 | 100 | 215 | 140 | 25 | 24 | 55 | 55 | 220 |
| AA6063 | 0 | 50 | 100 | 70 | 27 | 26 | 25 | 85 | 110 |
| | T1 | 90 | 150 | 95 | 26 | 24 | 45 | 45 | 150 |
| | T4 | 90 | 160 | 110 | 21 | 21 | 50 | 50 | 150 |
| | T5 | 175 | 215 | 135 | 14 | 13 | 60 | 65 | 150 |
| | T6 | 210 | 245 | 150 | 14 | 12 | 75 | 80 | 150 |
| | T8 | 240 | 260 | 155 | | 9 | 80 | 85 | |
| AA6082 | 0 | 60 | 130 | 85 | 27 | 26 | 35 | 35 | 120 |
| | T1 | 170 | 260 | 155 | 24 | 24 | 70 | 75 | 200 |
| | T4 | 170 | 260 | 170 | 19 | 19 | 70 | 75 | 200 |
| | T5 | 275 | 325 | 195 | 11 | 11 | 90 | 95 | 210 |
| | T6 | 310 | 340 | 210 | 11 | 11 | 95 | 100 | 210 |
| AA6262 | T6 | 240 | 290 | | 8 | | | | |
| | T9 | 330 | 360 | | 3 | | | | |
| AA7075 | 0 | 105 | 225 | 150 | | 17 | 60 | 65 | 230 |
| | T6 | 505 | 570 | 350 | 10 | 10 | 150 | 160 | 300 |
| | T7 | 435 | 505 | 305 | 13 | 12 | 140 | 150 | 300 |

## Aluminium Standards

The old BS1470 standard has been replaced by nine EN standards. The EN standards are given in table.

### Table. EN standards for aluminium

| Standard | Scope |
|----------|-------|
| EN485-1 | Technical conditions for inspection and delivery |
| EN485-2 | Mechanical properties |
| EN485-3 | Tolerances for hot rolled material |
| EN485-4 | Tolerances for cold rolled material |
| EN515 | Temper designations |
| EN573-1 | Numerical alloy designation system |
| EN573-2 | Chemical symbol designation system |
| EN573-3 | Chemical compositions |
| EN573-4 | Product forms in different alloys |

The EN standards differ from the old standard, BS1470 in the following areas:

- Chemical compositions – unchanged.

- Alloy numbering system – unchanged.

- Temper designations for heat treatable alloys now cover a wider range of special tempers. Up to four digits after the T have been introduced for non- standard applications (e.g. T6151).

- Temper designations for non-heat treatable alloys – existing tempers are unchanged but tempers are now more comprehensively defined in terms of how they are created. Soft (O) temper is now H111 and an intermediate temper H112 have been introduced. For alloy 5251 tempers are now shown as H32/H34/H36/H38 (equivalent to H22/H24, etc). H19/H22 & H24 are now shown separately.

- Mechanical properties – remain similar to previous figures. 0.2% Proof Stress must now be quoted on test certificates.

- Tolerances have been tightened to various degrees.

## Heat Treatment of Aluminium

A range of heat treatments can be applied to aluminium alloys:

- Homogenisation – the removal of segregation by heating after casting.

- Annealing – used after cold working to soften work-hardening alloys (1XXX, 3XXX and 5XXX).

- Precipitation or age hardening (alloys 2XXX, 6XXX and 7XXX).

- Solution heat treatment before ageing of precipitation hardening alloys.

- Stoving for the curing of coatings.

- After heat treatment a suffix is added to the designation numbers.

- The suffix F means "as fabricated".

- O means "annealed wrought products".

- T means that it has been "heat treated".

- W means the material has been solution heat treated.

- H refers to non-heat treatable alloys that are "cold worked" or "strain hardened".

The non-heat treatable alloys are those in the 3XXX, 4XXX and 5XXX groups.

Table. Heat treatment designations for aluminium and aluminium alloys.

| Term | Description |
|------|-------------|
| T1 | Cooled from an elevated temperature shaping process and naturally aged. |
| T2 | Cooled from an elevated temperature shaping process cold worked and naturally aged. |
| T3 | Solution heat-treated cold worked and naturally aged to a substantially. |
| T4 | Solution heat-treated and naturally aged to a substantially stable condition. |
| T5 | Cooled from an elevated temperature shaping process and then artificially aged. |
| T6 | Solution heat-treated and then artificially aged. |
| T7 | Solution heat-treated and overaged/stabilised. |

## Work Hardening of Aluminium

The non-heat treatable alloys can have their properties adjusted by cold working. Cold rolling is an example.

These adjusted properties depend upon the degree of cold work and whether working is followed by any annealing or stabilising thermal treatment.

Nomenclature to describe these treatments uses a letter, O, F or H followed by one or more numbers. The first number refers to the worked condition and the second number the degree of tempering.

There are a whole host of reasons why aluminium is so suitable as a building material. One of aluminium's most important advantages, no matter what the industry, is its high strength to weight ratio. In building construction, this characteristic has many specific advantages. For example, aluminium alloys in the infrastructure of a building are able to bear the weight of heavy glass spans, an important consideration in many modern office buildings and skyscrapers. The ability to use so much glass allows for more sunlight, thus reducing lighting costs while also saving on heating and air conditioning.

Another benefit of using aluminium is its air tightness. The cracks that can naturally occur in materials in the window frames are a major source of air leakage, which can increase the heating and cooling load of buildings. Aluminium is known for its air tightness and does not suffer the same deterioration that plagues materials such as steel and wood. For example, in the 90's, the steel frames of the Empire State Building were replaced by newer aluminium frames to help increase the landmark's energy efficiency.

This leads to the next major advantage of aluminium over rival materials, which is its durability. Aluminium is known for its high corrosion resistance. Anodized aluminium is both highly receptive to polishing and incredibly long lasting, an important consideration for the construction industry, as long term maintenance costs will be much lower than comparable materials. Aluminium also isn't susceptible to weathering conditions and can withstand in both humid and dry climates extremely well, nor does it become brittle in low temperatures.

Aluminium alloys are also highly valued for their appearance. Because of aluminium's receptiveness to being polished, it's possible to get smooth and bright finishing that is aesthetically pleasing. Aluminium can also take on any color following the anodizing process by immersing it in a warm bath of colouring agents, making it a favourite for designers to work with. In fact, one of the earliest uses of aluminium in the building process was for decorative purposes.

Because of its versatility and light weight, aluminium is also much easier to work with than most other materials. This leads to a great deal of cost savings during the construction process. Aluminium alloys can be welded, cast, forged, extruded, or rolled, meaning that aluminium can be found throughout the building, from its main support structure to the decorative elements and electrical fixtures. Another added benefit is that because of its light weight, it's easier to transport, an important cost factor during construction.

With environmental concerns becoming more and more prominent, architects and building owners have made finding green materials that can contribute to a building's overall sustainability an important concern. The use of aluminium alloys has helped many building projects qualify for the Leadership in Energy and Environmental Design (LEED) standards.

Because of aluminium's high reflexivity, buildings that use aluminium can reflect heat and save on cooling costs. As mentioned, aluminium is strong enough to hold large glass panels as well, another way of regulating the temperature of a building. By reducing the amount of heating and cooling required, this also significantly lowers the overall carbon footprint of a building over its lifetime.

Another bonus of aluminium is that it's 100% recyclable, losing none of its properties during the recycling process. By using recycled aluminium, the energy consumption is reduced by 90% compared with having to produce new aluminium. And because of its adaptability and form-ability, aluminium is often used to refurbish older buildings, making them more energy efficient and longer lasting.

The alloys most widely used in the construction of buildings and other large structures come from the 5000 and 6000 series, with the latter being most popular. This is due to it being highly extrudable and therefore capable of being shaped with greater versatility. Silicone alloys (ex: LM6) and manganese alloys (ex: 3103) can also be employed in certain building applications.

When selecting an alloy, all of its properties must be taken into account. The 6063 alloy is valued for its combination of extrudability, corrosion resistance, and surface finish, making it popular in window frames. Individual alloys can also be affected by the shape of the casting die. Builders and architects need to be very knowledgeable about how alloys will respond to particular shapes and whether their performance and durability will be affected.

# References

- Kohl, Walter H. (1995). Handbook of materials and techniques for vacuum devices. Springer. pp. 164–167. ISBN 1-56396-387-6

- Wrought-iron-properties-and-uses-metals-industries-metallurgy-20663: engineeringenotes.com, Retrieved 06 May 2018

- Boehler, Reinhard (2000). "High-pressure experiments and the phase diagram of lower mantle and core materials". Review of Geophysics. American Geophysical Union. 38 (2): 221–245. Bibcode:2000RvGeo..38..221B. doi:10.1029/1998RG000053

- Cast-iron-types-properties-uses-16022: theconstructor.org, Retrieved 24 April 2018

- Tewari, Rakesh. "The origins of Iron Working in India: New evidence from the Central Ganga plain and the Eastern Vindhyas" (PDF). State Archaeological Department. Retrieved 23 May 2010

- What-is-structural-steel-and-how-is-it-created: swantonweld.com, Retrieved 14 May 2018

- Witzel, Michael (2001), "Autochthonous Aryans? The Evidence from Old Indian and Iranian Texts", in Electronic Journal of Vedic Studies (EJVS) 7-3, pp. 1–93

- Properties-structural-steel-design-construction-18932: theconstructor.org, Retrieved 24 June 2018

- Kuhn, Howard and Medlin, Dana (prepared under the direction of the ASM International Handbook Committee), eds. (2000). ASM Handbook – Mechanical Testing and Evaluation (PDF). 8. ASM International. p. 275. ISBN 0-87170-389-0

- Cement-concrete-applications: cement.org, Retrieved 11 March 2018

- Stixrude, Lars; Wasserman, Evgeny; Cohen, Ronald E. (1997-11-10). "Composition and temperature of Earth's inner core". Journal of Geophysical Research: Solid Earth. 102 (B11): 24729–24739. Bibcode:1997JGR...10224729S. doi:10.1029/97JB02125

- Properties-reinforced-concrete: merloconstructionmi.com, Retrieved 14 April 2018

- Rokni, Sayed H.; Cossairt, J. Donald; Liu, James C. (January 2008). "Radiation Shielding at High-Energy Electron and Proton Accelerators" (PDF). Retrieved 6 August 2016

- Aluminum-uses-in-construction: clintonaluminum.com, Retrieved 24 May 2018

- Whitaker, Robert D (1975). "An historical note on the conservation of mass". Journal of Chemical Education. 52 (10): 658. Bibcode:1975JChEd..52..658W. doi:10.1021/ed052p658

# 4

# Strength and Stiffness

An understanding of how structures resist and support imposed loads and self-weight is vital to structural engneering. Structural analysis, theoretical and empirical design codes and corrosion resistance of structures and materials are other aspects intrinsic to this field. The topics elucidated in this chapter on stiffness, bending stiffness, deflection, strength analysis, etc. aid an extensive understanding of structural engineering.

## Strength Analysis

### Strength of Materials

Strength of materials is also known as mechanics of materials. It is focused on analysing stresses and deflections in materials under load. Knowledge of stresses and deflections allows for the safe design of structures that are capable of supporting their intended loads.

### Stress & Strain

When a force is applied to a structural member, that member will develop both stress and strain as a result of the force. Stress is the force carried by the member per unit area, and typical units are lbf/in² (psi) for US Customary units and N/m² (Pa) for SI units:

$$\sigma = \frac{F}{A}$$

where F is the applied force and A is the cross-sectional area over which the force acts. The applied force will cause the structural member to deform by some length, in proportion to its stiffness. Strain is the ratio of the deformation to the original length of the part:

$$\epsilon = \frac{L - L_o}{L_o} = \frac{\delta}{L_o}$$

where L is the deformed length, $L_o$ is the original undeformed length, and $\delta$ is the deformation (the difference between the two).

There are different types of loading which result in different types of stress, as outlined in the table below:

| Loading Type | Stress Type | Illustration |
|---|---|---|
| Axial Force | • Axial Stress (general case)<br><br>• Tensile Stress (if force is tensile)<br><br>• Compressive Stress (if force is compressive) | |
| Shear Force | Transverse Shear Stress | |
| Bending Moment | Bending Stress | |
| Torsion | Torsional Stress | |

Axial stress and bending stress are both forms of normal stress, $\sigma$, since the direction of the force is normal to the area resisting the force. Transverse shear stress and torsional stress are both forms of shear stress, $\tau$, since the direction of the force is parallel to the area resisting the force.

| Normal Stress | |
|---|---|
| Axial Stress: | $\sigma = \dfrac{F}{A}$ |
| Bending Stress: | $\sigma_b = \dfrac{My}{I_c}$ |

| Shear Stress | |
|---|---|
| Transverse Stress: | $\tau = \dfrac{F}{A}$ |
| Torsional Stress: | $\tau = \dfrac{Tr}{J}$ |

In the equations for axial stress and transverse shear stress, $F$ is the force and $A$ is the cross-sectional area of the member. In the equation for bending stress, $M$ is the bending moment, $y$ is the distance between the centroidal axis and the outer surface, and $I_c$ is the centroidal moment of inertia of the cross section about the appropriate axis. In the equation for torsional stress, $T$ is the torsion, $r$ is the radius, and $J$ is the polar moment of inertia of the cross section.

In the case of axial stress over a straight section, the stress is distributed uniformly over the entire area. In the case of shear stress, the distribution is maximum at the center of the cross section; however, the average stress is given by $\tau = F/A$, and this average shear stress is commonly used in stress calculations. More discussion can be found in the section on shear stresses in beams. In the case of bending stress and torsional stress, the maximum stress occurs at the outer surface. More discussion can be found in the section on bending stresses in beams.

Just as the primary types of stress are normal and shear stress, the primary types of strain are normal strain and shear strain. In the case of normal strain, the deformation is normal to the area carrying the force:

$$\grave{o} = \frac{\delta}{L_o} = \frac{\sigma}{E}$$

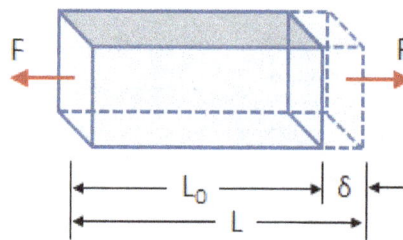

In the case of transverse shear strain, the deformation is parallel to the area carrying the force:

$$\gamma = \frac{\delta_{parallel\ to\ area}}{H} = \tan\phi \approx \phi$$

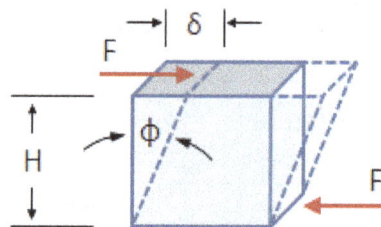

where $\gamma$ is the shear strain (unitless) and $\phi$ is the deformed angle in units of radians.

In the case of torsional strain, the member twists by an angle, $\phi$, about its axis. The maximum shear strain occurs on the outer surface. In the case of a round bar, the maximum shear strain is given by:

$$\gamma_{max} = \frac{r\phi}{L}$$

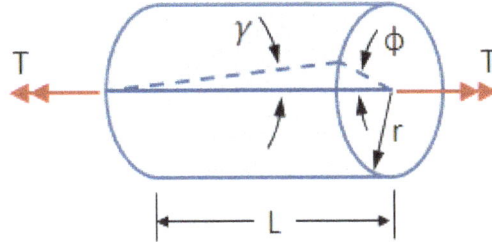

where $\phi$ is the angle of twist, $r$ is the radius of the bar and $L$ is the length.

The shear strains are proportional through the interior of the bar, and are related to the max shear strain at the surface by:

$$\gamma = \frac{\rho}{r}\gamma_{max}$$

where $\rho$ is the radial distance from the bar's axis.

## Hooke's Law

Stress is proportional to strain in the elastic region of the material's stress-strain curve (below the proportionality limit, where the curve is linear).

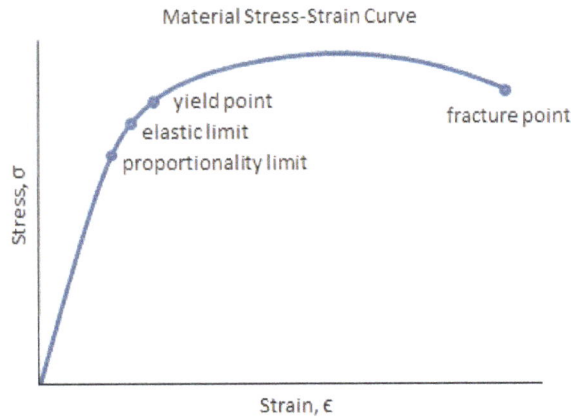

Material Stress-Strain Curve

Normal stress and strain are related by:

$$\sigma = E\dot{o}$$

where $E$ is the elastic modulus of the material, $\sigma$ is the normal stress, and $\dot{o}$ is the normal strain.

Shear stress and strain are related by:

$$\tau = G\gamma$$

where $G$ is the shear modulus of the material, $\tau$ is the shear stress, and $\gamma$ is the shear strain. The elastic modulus and the shear modulus are related by:

$$G = \frac{E}{2(1+\nu)}$$

where $\nu$ is Poisson's ratio.

Hooke's law is analogous to the spring force equation, $F = k\delta$ Essentially, everything can be treated like a spring. Hooke's Law can be rearranged to give the deformation (elongation) in the material:

| Axial Elongation (from normal stress) | $\delta = L_o \epsilon = \dfrac{L_o \sigma}{E} = \dfrac{FL_o}{AE}$ |  |
|---|---|---|
| Angle of Twist (from shear/torsional stress) | $\phi = \dfrac{\tau L}{rG} = \dfrac{TL}{GJ}$ |  |

## Strain Energy

When force is applied to a structural member, that member deforms and stores potential energy, just like a spring. The strain energy (i.e. the amount of potential energy stored due to the deformation) is equal to the work expended in deforming the member. The total strain energy corresponds to the area under the load deflection curve, and has units of in-lbf in US Customary units and N-m in SI units. The elastic strain energy can be recovered, so if the deformation remains within the elastic limit, then all of the strain energy can be recovered.

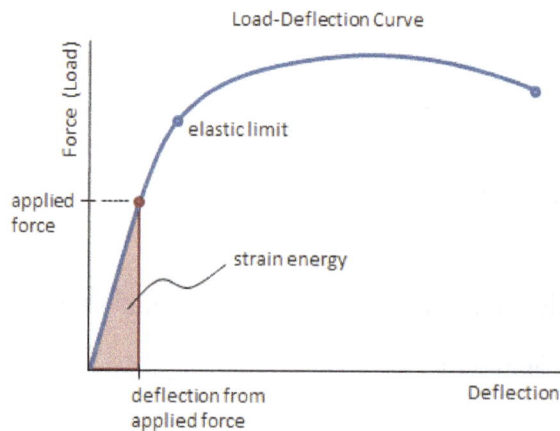

Strain energy is calculated as:

| General Form: | $U = Work = \int F\ dL$ | (area under load-deflection curve) |
|---|---|---|
| Within Elastic Limit: | $U = \dfrac{1}{2}F\delta = \dfrac{F^2 L_o}{2AE} = \dfrac{\sigma^2 L_o A}{2E}$ | (area under load-deflection curve) |
| | $U = \dfrac{1}{2}k\delta^2 = \dfrac{AE\delta^2}{2L_o}$ | (spring potential energy) |

Note that there are two equations for strain energy within the elastic limit. The first equation is based on the area under the load deflection curve. The second equation is based on the equation for the potential energy stored in a spring. Both equations give the same result, they are just derived somewhat differently.

## Need Structural Calculators

We have a number of structural calculators to choose from. Here are just a few:

- Beam Calculator
- Bolted Joint Calculator
- Bolt Pattern Force Distribution
- Lug Calculator
- Column Buckling Calculator

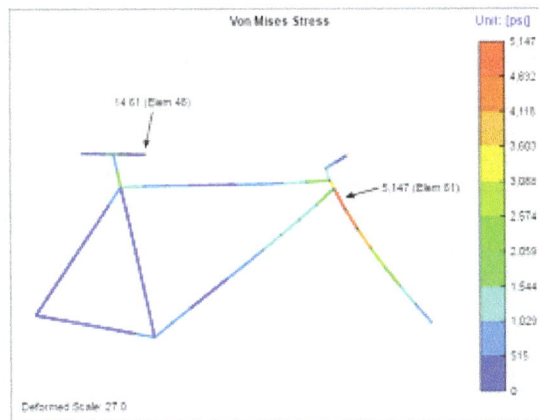

## Stiffness

Stiffness, commonly referred to as the spring constant, is the force required to deform a structural member by a unit length. All structures can be treated as collections of springs, and the forces and deformations in the structure are related by the spring equation:

$$F = k\delta_{max}$$

where $k$ is the stiffness, $F$ is the applied force, and $\delta_{max}$ is the maximum deflection deflection in the member.

If the deflection is known, then the stiffness of the member can be found by solving $k = F / \delta_{max}$.

However, the maximum deflection is typically not known, and so the stiffness must be calculated by other means. Beam deflection tables can be used for common cases. The two most useful stiffness equations to know are those for a beam with an axially applied load, and for a cantilever beam with an end load. Note that stiffness is a function of the material's elastic modulus, $E$, the geometry of the part, and the loading configuration.

| | Stiffness [lbf/in] | Max Deflection [in] | Illustration |
|---|---|---|---|
| Beam with axially applied load: | $k = \dfrac{AE}{L_o}$ | $\delta = \dfrac{FL_o}{AE}$ | |
| Cantilever beam with end load: | $k = \dfrac{3EI}{L^3}$ | $\delta_{max} = \dfrac{FL^3}{3EI}$ | |

The torsional equivalent of the of the spring equation is:

$$T = k\phi$$

The stiffness of a shaft under torsional load is of particular interest:

| | Stiffness [in*lbf/rad] | Max Deflection [rad] | Illustration |
|---|---|---|---|
| Shaft with torsional load: | $k = \dfrac{GJ}{L}$ | $\phi = \dfrac{TL}{GJ}$ | |

## Structure with Multiple Load Paths

If there are multiple load paths in a structure (i.e. there are multiple members in a structure that share the load), the load will be higher in the stiffer members. To find the load carried by any individual member, first calculate the equivalent stiffness of the members in the load path, treating them as springs. Depending on their configuration, they will be treated as some combination of springs in series and springs in parallel.

| | Springs in Parallel | Springs in Series |
|---|---|---|
| Illustration: | 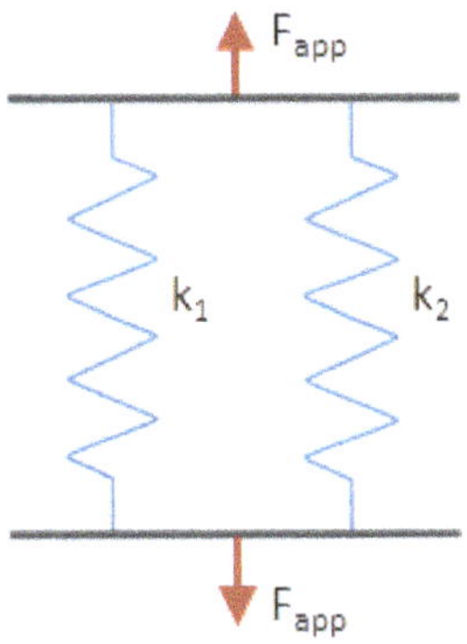 | 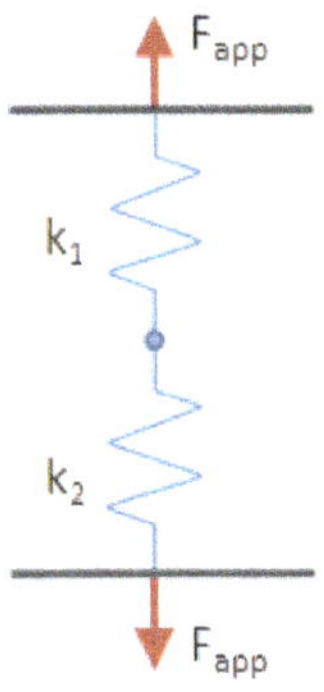 |
| Equivalent Stiffness of Structure: | $k_{eq} = k_1 + k_2 + k_3 + \dots$ | $keq = \dfrac{1}{1/k1 + 1/k2 + 1/k3 + \dots}$ |
| Force in Individual Member: | $F_1 = \dfrac{k_1}{k_{eq}} F_{app}$ <br><br> $F_1 = \dfrac{k_1}{k_2} F_2$ | $F_1 = k_{eq}\delta_{tot} = F_{app}$ |
| Deflection in Individual Member: | $\delta_1 = \dfrac{F_1}{k_1} = \dfrac{F_{app}}{k_{eq}}$ | $\delta_1 = \dfrac{k_{eq}}{k_1}\delta_{tot}$ <br><br> $\delta_1 = \dfrac{k_2}{k_1}\delta_2$ |
| Total Force in Structure: | $F_{tot} = F_1 + F_2 + F_3 + \dots$ | $F_{tot} = F_1 = F_2 = F_3 = \dots$ |
| Total Deflection in Structure: | $\delta_{tot} = \delta_1 = \delta_2 = \delta_3 = \dots$ | $\delta_{tot} = \delta_1 + \delta_2 = \delta_3 = \dots$ |

If the members in the load path cannot be treated purely as springs in series or as springs in parallel, but are rather a combination of springs in series and in parallel, then the problem will need to be solved iteratively. Fine a sub-grouping of the members that are either purely in series or in parallel and use the equations provided to calculate the equivalent stiffness, force and deflection in the sub-group. The sub-group can then be considered a single spring with the calculated stiffness, force, and deflection, and that spring can then be considered as a part of another sub-group of springs. Continue grouping members and solving until the desired result is achieved.

## Stress Concentrations

Forces and stresses can be thought to flow through a material, as shown in the figure below. When the geometry of the material changes, the flow lines move closer together or farther apart to accommodate. If there is a discontinuity in the material such as a hole or a notch, the stress must flow around the discontinuity, and the flow lines will pack together in the vicinity of that discontinuity. This sudden packing together of the flow lines causes the stress to spike up this peak stress is called a stress concentration. The feature that causes the stress concentration is called a stress riser.

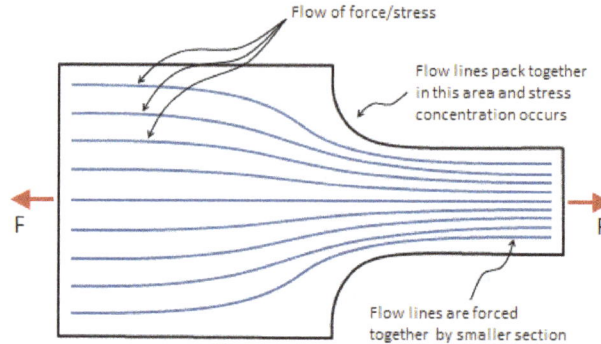

Stress concentrations are accounted for by stress concentration factors. To find the actual stress in the vicinity of a discontinuity, calculate the nominal stress in that area and then scale it up by the appropriate stress concentration factor:

$$\sigma_{max} = K\sigma_{nom}$$

where $\sigma_{max}$ is the actual (scaled) stress, $\sigma_{nom}$ is the nominal stress, and $K$ is the stress concentration factor. When calculating the nominal stress, use the maximum value of stress in that area. For example, in the figure above, the smallest area at the base of the fillet should be used.

Many reference handbooks contain tables and curves of stress concentration factors for various geometries. Two of the most comprehensive collections of stress concentration factors are Peterson's Stress Concentration Factors and Roark's Formulas for Stress and Strain. Mechanical also provides a collection of interactive plots for common stress concentration factors.

The concentration of stress will dissipate as we move away from the stress riser. Saint-Venant's principle is a general rule of thumb stating that the distance over which the stress concentration dissipates is equal to the largest dimension of the cross section carrying the load.

Calculation of stress concentration is particularly important when the materials are very brittle, or when there is only a single load path. In ductile materials, local yielding will allow for stresses to be redistributed and will reduce the stress around the riser. For this reason, stress concentration factors are not typically applied to structural members made of ductile materials. Stress concentration factors are also not typically applied when there is a redundant load path, in which case yielding of one member will allow for redistribution of forces to the members on the other load paths. An example of this is a pattern of bolts. If one bolt starts to give, then the other bolts in the pattern will take more of the load.

## Combined Stresses

At any point in a loaded material, a general state of stress can be described by three normal stresses (one in each direction) and six shear stresses (two in each direction):

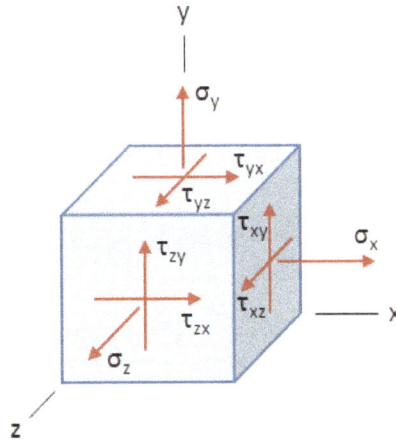

The subscripts on the normal stresses, $\sigma$, indicate the direction of the normal stress. The subscripts of the shear stresses, $\tau$, have two components. The first indicates the direction of the surface normal, and the second indicates the direction of the shear stress itself.

Commonly, the stresses along one direction are zero so that the full state of stress occurs on a single plane, as shown in the figure below. This is called plane stress. Plane stress occurs in thin plates, but it also occurs on the surface of any loaded structure. Surface stresses are commonly the most critical stresses since bending stress and torsional stress are maximized at the surface.

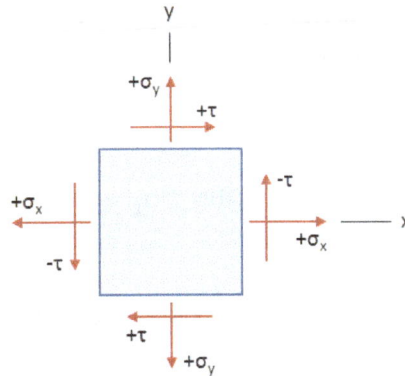

In the figure above, $\sigma_x$ and $\sigma_y$ are the normal stresses, and $\tau$ is the shear stress. The stresses balance so that the point is in static equilibrium. Because the shear stresses are all equal in magnitude, the subscripts are dropped for simplicity. (Note however that the sign of the stresses on the $x$ face will be opposite to those on the $y$ face.)

The proper sign conventions are as shown in the figure. For normal stress, tensile stress is positive and compressive stress is negative. For shear stress, clockwise is positive and counter clockwise is negative.

If the stresses from the figure above are known, it is possible to find the normal and shear stress on a plane rotated at some angle, $\theta$, with respect to the horizontal, as shown in the figure below. The transformation equations below give the values of the normal stress and shear stress on this rotated plane.

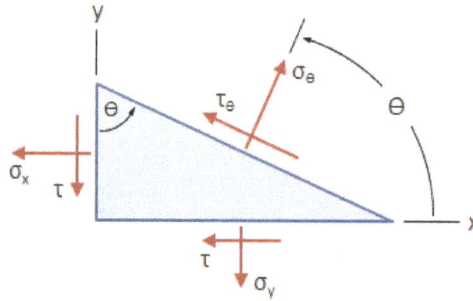

| Normal Stress: | $\sigma_\theta = \dfrac{\sigma_x + \sigma_y}{2} + \dfrac{\sigma_x - \sigma_y}{2}\cos 2\theta + \tau \sin 2\theta$ |
|---|---|
| Shear Stress: | $\tau_\theta = -\dfrac{\sigma_x - \sigma_y}{2}\sin 2\theta + \tau \cos 2\theta$ |

Note that in the figure above, $\theta$ is measured from the x-axis, and a positive value of $\theta$ is counter clockwise.

At any point in the material, it is possible to find the angles of the plane at which the normal stresses and the shear stresses are maximized and minimized. The maximum and minimum normal stresses are called principal stresses. The maximum and minimum shear stresses are called the extreme shear stresses. The angles of the principal stresses and the extreme shear stresses are found by taking the derivative of each transformation equation with respect to $\theta$ and finding the value of $\theta$ where the derivative is zero.

| Principal Stress Angles: | $\theta\sigma_1, \theta\sigma_2 = \dfrac{1}{2}arctan\left(\dfrac{2\tau}{\sigma x - \sigma y}\right)$ |
|---|---|
| Extreme Shear Stress Angles: | $\theta\tau_1, \theta\tau_2 = \dfrac{1}{2}arctan(\sigma x - \sigma y - 2\tau)$ |

The angles above can be substituted back into the transformation equations to find the values of the principal stresses and the extreme shear stresses:

| Principal Stresses: | $\sigma_1, \sigma_2 = \dfrac{\sigma_x + \sigma_y}{2} \pm \sqrt{\left(\dfrac{\sigma_x - \sigma_y}{2}\right)^2 + \tau^2}$ |
|---|---|
| Extreme Shear Stresses: | $\tau_1, \tau_2 = \pm\sqrt{\left(\dfrac{\sigma_x - \sigma_y}{2}\right)^2 + \tau_2}$ |

The angles at which the principal stresses occur are 90° apart. Principal stresses are always accompanied by zero shear stress. The angles at which the extreme shear stresses occur are 45° from the angles of the principal stresses. Extreme shear stresses are accompanied by two equal normal stresses of $(\sigma_x + \sigma_y)/2$.

A couple useful relationships are:

| | |
|---|---|
| $\sigma_1 + \sigma_2 = \sigma_x + \sigma_y$ | The sum of the normal stresses is constant. |
| $\tau_1 = \dfrac{\sigma_1 - \sigma_2}{2}$ | The maximum shear stress is half the difference of the principal stresses. |

## Mohr's Circle

Mohr's circle is a way of visualizing the state of stress at a point on a loaded material. It gives an intuitive feel to the stress transformation equations, and shows how the stresses on an element change as a function of the rotation angle, $\theta$. From Mohr's circle, it also becomes clear what are the principal stresses, the extreme shear stresses, and the angles at which those stresses occur. An example of Mohr's circle is shown in the figure below:

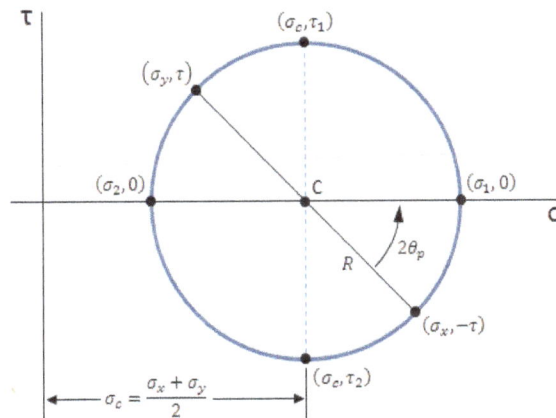

To construct Mohr's circle, first locate the centre of the circle by taking the average of the normal stresses:

$$\sigma_c = \frac{\sigma_x + \sigma_y}{2}$$

Place points on the circle representing the stresses on the $x$ and $y$ faces of the stress element. The stresses on the $x$ face will have the coordinates $(\sigma_x, -\tau)$, and the stresses on the $y$ face will have the coordinates $(\sigma_y, \tau)$. Place points on the circle for the principal stresses. The maximum principal stress will have the coordinates $(\sigma_1, 0)$, and the minimum principal stress will have the coordinates $(\sigma_2, 0)$. Place points on the circle for the extreme shear stresses. The maximum extreme shear stress will have coordinates $(\sigma_c, \tau_1)$, and the minimum extreme shear stress will have coordinates $(\sigma_c, \tau_2)$.

All of the points will lie on the perimeter of the circle. The circle has a radius equal to the magnitude of the extreme shear stresses:

$$R = \sqrt{\left(\frac{\sigma_x - \sigma_y}{2}\right)^2 + \tau^2}$$

The state of stress on the $x$ and $y$ faces of the stress element is represented by the black line in Mohr's circle connecting the points $(\sigma_x, -\tau)$ and $(\sigma_y, \tau)$. This line in Mohr's circle corresponds to the unrotated element in the figure below. If this line is rotated by some angle, then the values of the points at the end of the rotated line will give the values of stress on the $x$ and $y$ faces of the rotated element. It is important to note that the 360 degrees of Mohr's circle are equivalent to 180 degrees on the stress element. For instance, the points for $x$ face and the $y$ face are 180 degrees apart on Mohr's circle, but they are only 90 degrees apart on the stress element.

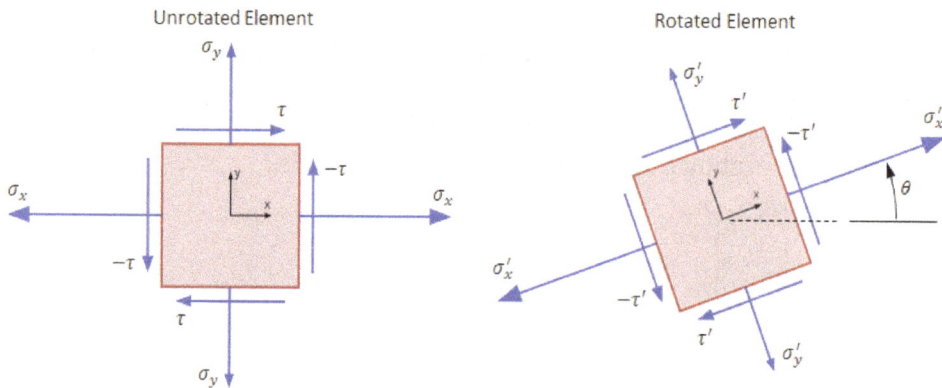

To get a more intuitive feel for how Mohr's circle relates the stresses on a stress element and how the stress state changes as a function of rotation angle, see the accompanying Mohr's circle calculator.

## Applications

There are many structural components that are commonly subjected to stress analysis. The details on the analysis of these components are given in other sections:

- Beams
- Columns
- Bolted joints
- Lifting lugs
- Pressure vessels
- Shafts

## Ultimate Tensile Strength

Ultimate tensile strength (UTS) is the maximum stress that a material can withstand while being stretched or pulled. The ultimate tensile strength of a material is calculated by dividing the cross-section area of the material tested by the stress placed on the material, generally expressed in terms of pounds or tons per square inch of material.

UTS is the final amount of stress sustained in a tensile test at the exact moment the object ruptures.

Tensile strength is an important measure of a material's ability to perform in an application, and the measurement is widely used when describing the properties of metals and alloys.

Ultimate tensile strength is also known as tensile strength or the ultimate.

The ultimate tensile strength (UTS) is a material's maximum resistance to fracture. It is equivalent to the maximum load that can be carried by one square inch of cross-sectional area when the load is applied as simple tension.

The UTS is the maximum engineering stress in a uniaxial stress-strain test. The UTS can differ, depending on the type of material:

- For non-deformable materials, it is the nominal stress at which a round bar of the material, loaded in tension, separates.

- For deformable materials, it occurs at the onset of necking at strains preceding breakage (separation).

- For brittle solids such as ceramics, glasses, and brittle polymers, it is the same as the failure strength in tension.

- For metals and most composites, it is larger than the yield strength by a factor of between 1.1 and 5 because of work hardening or, in the case of composites, load transfer to the re-inforcement.

The UTS is not used in the design of ductile static materials because design practices dictate the use of the yield stress. It is, however, used for quality control, because of the ease of testing. It is a common engineering parameter when designing brittle materials, because there is no yield point.

The UTS is usually found by performing a tensile test and recording the engineering stress versus strain curve. The highest point of the stress-strain curve is the UTS. It is an intensive property; therefore its value does not depend on the size of the test specimen. However, it is dependent on other factors, such as:

- Preparation of the specimen.

- Presence of surface defects.

- Temperature of the test environment and the material.

## Ductile Materials

Many materials can display linear elastic behavior, defined by a linear stress–strain relationship, as shown in figure 1 up to point 3. The elastic behavior of materials often extends into a non-linear region, represented in figure 1 by point 2 (the "yield point"), up to which deformations are completely recoverable upon removal of the load; that is, a specimen loaded elastically in tension will elongate, but will return to its original shape and size when unloaded. Beyond this elastic region, for ductile materials, such as steel, deformations are plastic. A plastically deformed specimen does not completely return to its original size and shape when unloaded. For many applications, plastic deformation is unacceptable, and is used as the design limitation.

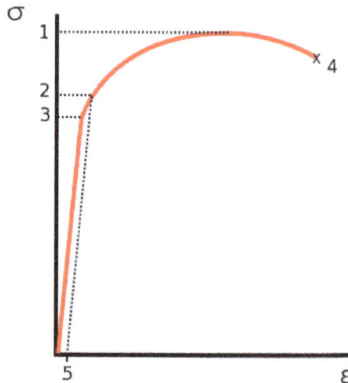

Figure: "Engineering" stress–strain (σ–ε) curve typical of aluminum 1. Ultimate strength 2. Yield strength 3. Proportional limit stress 4. Fracture 5. Offset strain (typically 0.2%)

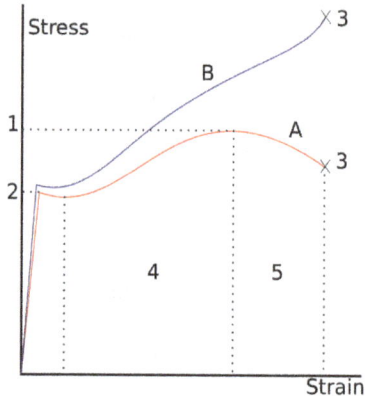

Figure: "Engineering" (red) and "true" (blue) stress–strain curve typical of structural steel. 1: Ultimate strength 2: Yield strength (yield point) 3: Rupture 4: Strain hardening region 5: Necking region A: Apparent stress (F/A$_o$) B: Actual stress (F/A)

After the yield point, ductile metals undergo a period of strain hardening, in which the stress increases again with increasing strain, and they begin to neck, as the cross-sectional area of the specimen decreases due to plastic flow. In a sufficiently ductile material, when necking becomes substantial, it causes a reversal of the engineering stress–strain curve (curve A, figure); this is because the *engineering stress* is calculated assuming the original cross-sectional area before necking. The reversal point is the maximum stress on the engineering stress–strain curve, and the engineering stress coordinate of this point is the ultimate tensile strength, given by point 1.

The UTS is not used in the design of ductile static members because design practices dictate the use of the yield stress. It is, however, used for quality control, because of the ease of testing. It is also used to roughly determine material types for unknown samples.

The UTS is a common engineering parameter to design members made of brittle material because such materials have no yield point.

## Testing

Typically, the testing involves taking a small sample with a fixed cross-sectional area, and then pulling it with a tensometer at a constant strain (change in gauge length divided by initial gauge length) rate until the sample breaks.

When testing some metals, indentation hardness correlates linearly with tensile strength. This important relation permits economically important nondestructive testing of bulk metal deliveries with lightweight, even portable equipment, such as hand-held Rockwell hardness testers. This practical correlation helps quality assurance in metalworking industries to extend well beyond the laboratory and universal testing machines.

While most metal forms, such as sheet, bar, tube, and wire, can exhibit the test UTS, fibers, such as carbon fibers, being only 2/10,000th of an inch in diameter, must be made into composites to create useful real-world forms. As the datasheet on T1000G below indicates, while the UTS of the fiber is very high at 6,370 MPa, the UTS of a derived composite is 3,040 MPa – less than half the strength of the fiber.

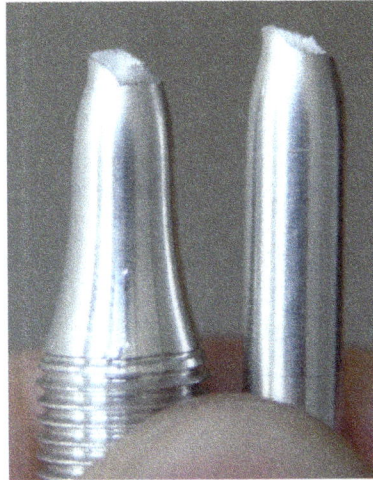

Round bar specimen after tensile stress testing

## Compressive Strength

Maximum stress a material can sustain under crush loading. The compressive strength of a material that fails by shattering fracture can be defined within fairly narrow limits as an independent property. However, the compressive strength of materials that do not shatter in compression must be defined as the amount of stress required to distort the material an arbitrary amount. Compressive strength is calculated by dividing the maximum load by the original cross-sectional area of a specimen in a compression test.

## Calculation of Compressive Strength

Compressive strength involves testing and calculating how well a given specimen, product or material can survive compressive stress. Unlike tension, which expands or pulls, compression means a specimen, product or material is shortened or pressed down. Compressive strength of a material is the point at which the material fails. Calculating compressive strength involves testing to find this failure point and using the data from the test to feed the calculation. The final compressive strength number is in pounds-force per square inch, or psi.

Set up a test to determine the maximum load of the specimen or material for which you are looking to calculate the compressive strength. This test has to be unique to your specimen. Determine the maximum amount of compressive stress the specimen can handle in units of pounds-force, or lbf. Place your specimen on a solid surface and apply a force to your specimen with the testing apparatus until the specimen fails or crushes. Upon failure, record the failure force in units of lbf. The force has to be adequate for the specimen to fail within 15 minutes of applying it.

Calculate the cross-section area, or A, of the specimen. If your specimen is circular or cylindrical, use the formula A = 3.1415r^2, where r is the radius of the specimen. If the specimen is four-sided, use the formula A = LW, where L is the length and W is the width. The area will be in units of in^2 or square inches. For example, if we assume you have a circular specimen with a radius of 2 inches, your area will be 12.56 in^2:

$$A = 3.1415r^2 = (3.1415)(2^2) = (3.1415)(4) = 12.56 \text{ in}^2$$

Calculate the compressive strength using the formula S = P/A, where S is the compressive strength, P is the maximum load applied to the specimen and A is the area. Your final compressive strength value will be in units of lbf/in^2, which is the same as pressure or psi. For example, if your maximum load is 20,000 lbf and your A is 12.56in^2, your compressive strength is 1,592 lbf/in^2 or 1,592 psi:

$$S = P/A = 20,000 \text{ lbf}/12.56 \text{ in}^2 = 1,592 \text{ lbf/in}^2 = 1,592 \text{ psi}$$

## Portland Cement Compressive Strength

The most common strength test, compressive strength, is carried out on a 50 mm (2-inch) cement mortar test specimen. The test specimen is subjected to a compressive load (usually from a hydraulic machine) until failure. This loading sequence must take no less than 20 seconds and no more than 80 seconds. Table shows ASTM C 150 compressive strength specifications.

Table: ASTM C 150 Portland Cement Mortar Compressive Strength Specifications in MPa (psi)

| Curing Time | Portland Cement Type | | | | | | | |
|---|---|---|---|---|---|---|---|---|
| | I | IA | II | IIA | III | IIIA | IV | V |
| 1 day | - | - | - | - | 12.4 (1800) | 10.0 (1450) | - | - |
| 3 days | 12.4 (1800) | 10.0 (1450) | 10.3 (1500) | 8.3 (1200) | 24.1 (3500) | 19.3 (2800) | - | 8.3 (1200) |
| 7 days | 19.3 (2800) | 15.5 (2250) | 17.2 (2500) | 13.8 (2000) | - | - | 6.9 (1000) | 15.2 (2200) |
| 28 days | - | - | - | - | - | - | 17.2 (2500) | 20.7 (3000) |

Note: Type II and IIA requirements can be lowered if either an optional heat of hydration or chemical limit on the sum of C3S and C3A is specified

## Standard Test Methods

- AASHTO T 106 and ASTM C 109: Compressive Strength of Hydraulic Cement Mortars (Using 50-mm or 2-in. Cube Specimens)

- ASTM C 349: Compressive Strength of Hydraulic Cement Mortars (Using Portions of Prisms Broken in Flexure).

## Maximum Compressive Strength

Maximum compressive strength that can possibly be obtained is generally reckoned to be about 7,500 psi for a 28-day cylinder. It would be interesting to know if there is a conceivable maximum limit using materials and techniques available. The limit at the moment seems to be 16,000 psi, although this can be obtained only under laboratory conditions. Using crushed rock or granite aggregates in an aggregate/cement ratio of 3:1 and pressure compaction, 6-inch cubes with a 14,000 psi strength at between 28 and 36 days can be obtained without any great difficulty. It would appear that the achievement of maximum density depends on the development of some radically new type of vibrator, possibly ultrasonic.

## Shear Strength

Shear strength is the strength of a material to withstand the type of yield or structural failure where the material or component fails in shear.

A shear load is a force that may produce a sliding failure on material along a plane that is parallel to the direction of the force. Shear strength is related to the type of fill material, whether it is high quality or poor quality, reflected in its stress and porosity.

## High Quality Fill

High quality "suitable" fill is usually a granular material with a limited percentage of fines and stone-sized fragments. Factors affecting the shear strength of granular material are particle size distribution, particle shape, mineralogy, angularity, degree of compaction and stress level. Other factors such as the amount of organic content and sulphates can also be considered. When shear stress is applied to a sand mass, the material will change in volume. Dense sand will dilate and loose sand will contract. This is an important characteristic of granular sands and influences the dredging process as well as the final use.

## Poor Quality Fill

Sometimes poor quality fill is the only fill available and must be used. In these cases, the shear strength of poor quality fill material can be improved and increased in several ways to make it suitable for use in a reclamation project, for instance:

- By a spraying and spreading pontoon;
- By a surcharge in combination with prefabricated vertical drains;
- By a sandwich construction with alternating clay and sand layers;
- By the use of binding materials like cement.

The use of poor quality fill can have serious consequences including:

- Very slow dewatering which makes it impossible to drive vehicles over the area during construction;
- The material is sensitive to vibrations and may liquefy when heavy equipment is being used;
- The development of strength, while it can be done by various methods mentioned above, is difficult to predict and must be tested.

The shear strength of cohesive material can be tested directly with a field vane test in a Shear Strength Testing.

## Shear Strength Testing

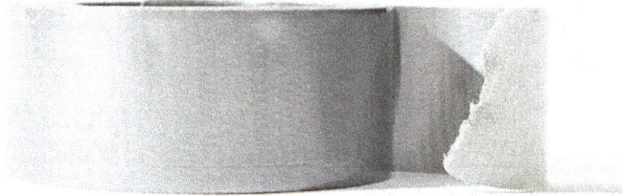

Shear testing is performed to determine the shear strength of a material. It measures the maximum shear stress that may be sustained before a material will rupture. Shear is typically reported as MPa (psi) based on the area of the sheared edge.

Shear testing is commonly used with adhesives and can be used in either a tensile or comprehensive method.

Typical graph showing a shear strength test:

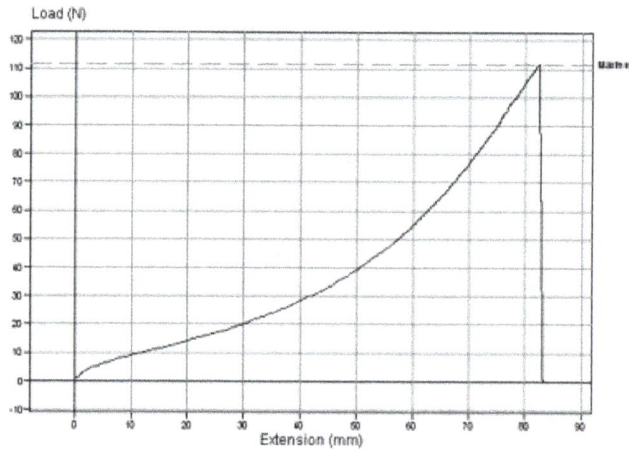

## Comparison

As a very rough guide relating tensile, yield, and shear strengths:

| Material | Ultimate Strength Relationship | Yield Strength Relationship |
|---|---|---|
| Steels | USS = approx. 0.75*UTS | SYS = approx. 0.58*TYS |
| Ductile Iron | USS = approx. 0.9*UTS | SYS = approx. 0.75*TYS . |
| Malleable Iron | USS = approx. 1.0*UTS | |
| Wrought Iron | USS = approx. 0.83*UTS | |
| Cast Iron | USS = approx. 1.3*UTS | |
| Aluminums | USS = approx. 0.65*UTS | SYS = approx. 0.55*TYS |

USS: Ultimate Shear Strength, UTS: Ultimate Tensile Strength, SYS: Shear Yield Stress, TYS: Tensile Yield Stress.

| Material | Ultimate stress (Ksi) | Ultimate stress (MPa) |
|---|---|---|
| Fiberglass/epoxy *(23 ° C)* | 7.82 | 53.9 |

When values measured from physical samples are desired, a number of testing standards are available, covering different material categories and testing conditions. In the US, ASTM standards for measuring shear strength include ASTM B831, D732, D4255, D5379, and D7078. Internationally, ISO testing standards for shear strength include ISO 3597, 12579, and 14130.

## Tensile Testing

A tensile test is also known as tension test. It is probably the most fundamental type of mechanical test you can perform on material. Tensile tests are simple, relatively inexpensive, and fully standardized. By pulling on something, you will very quickly determine how the material will react to forces being applied in tension. As the material is being pulled, you will find its strength along with how much it will elongate.

## Reason for Performing Tensile Tests

You can learn a lot about a substance from tensile testing. As you continue to pull on the material until it breaks, you will obtain a good, complete tensile profile. A curve will result showing how it reacted to the forces being applied. The point of failure is of much interest and is typically called its "Ultimate Strength" or UTS on the chart.

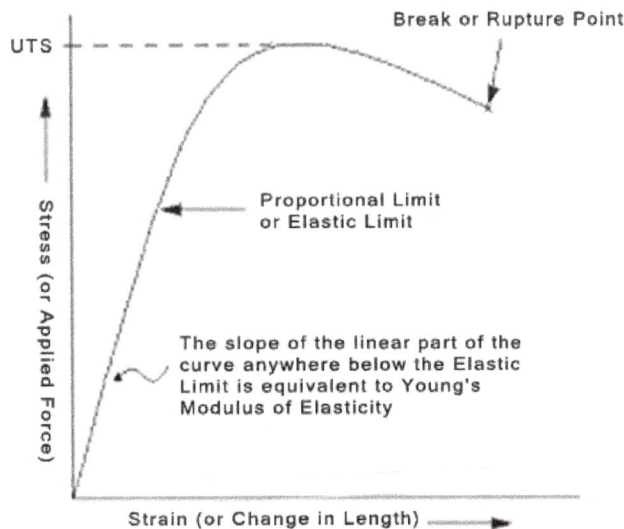

## Hooke's Law

For most tensile testing of materials, you will notice that in the initial portion of the test, the relationship between the applied force, or load, and the elongation the specimen exhibits is linear.

In this linear region, the line obeys the relationship defined as "Hooke's Law" where the ratio of stress to strain is a constant, or $\dfrac{\sigma}{\varepsilon} = E$. E is the slope of the line in this region where stress ($\sigma$) is proportional to strain ($\varepsilon$) and is called the "Modulus of Elasticity" or "Young's Modulus".

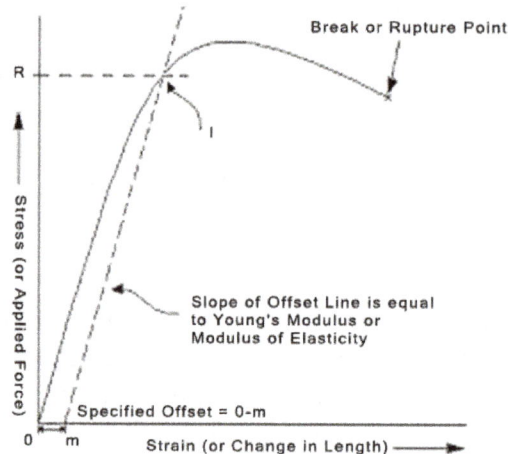

## Modulus of Elasticity

The modulus of elasticity is a measure of the stiffness of the material, but it only applies in the linear region of the curve. If a specimen is loaded within this linear region, the material will return to its exact same condition if the load is removed. At the point that the curve is no longer linear and deviates from the straight-line relationship, Hooke's Law no longer applies and some permanent deformation occurs in the specimen. This point is called the "elastic, or proportional limit". From this point on in the tensile test, the material reacts plastically to any further increase in load or stress. It will not return to its original, unstressed condition if the load were removed.

## Yield Strength

A value called "yield strength" of a material is defined as the stress applied to the material at which plastic deformation starts to occur while the material is loaded.

## Offset Method

For some materials (e.g., metals and plastics), the departure from the linear elastic region cannot be easily identified. Therefore, an offset method to determine the yield strength of the material tested is allowed. These methods are discussed in ASTM E8 (metals) and D638 (plastics). An offset is specified as a % of strain (for metals, usually 0.2% from E8 and sometimes for plastics a value of 2% is used). The stress (R) that is determined from the intersection point "r" when the line of the linear elastic region (with slope equal to Modulus of Elasticity) is drawn from the offset "m" becomes the Yield Strength by the offset method.

## Alternate Moduli

The tensile curves of some materials do not have a very well-defined linear region. In these cases,

ASTM Standard E111 provides for alternative methods for determining the modulus of a material, as well as Young's Modulus. These alternate moduli are the secant modulus and tangent modulus.

## Strain

You will also be able to find the amount of stretch or elongation the specimen undergoes during tensile testing This can be expressed as an absolute measurement in the change in length or as a relative measurement called "strain". Strain itself can be expressed in two different ways, as "engineering strain" and "true strain". Engineering strain is probably the easiest and the most common expression of strain used. It is the ratio of the change in length to the original length,

$e = \dfrac{L - L_o}{L_o} = \dfrac{\Delta L}{L_o}$. Whereas, the true strain is similar but based on the instantaneous length of

the specimen as the test progresses, $\varepsilon = In\left(\dfrac{L_i}{L_o}\right)$, where $L_i$ is the instantaneous length and $L_o$ the

initial length.

## Types of Tensile Tests

There many different variants of tensile tests but a few of the more common tests are tension, tensile adhesion, tensile shear, tensile grab, tensile pulling, tension fatigue, and tensile creep. In most of these tests the specimen is loaded until it fails or fractures with the main difference in the type of specimen geometry and associated tensile test fixture used. Tension fatigue testing differs not only in the type of grip but also in the test machine type. It is performed by loading the material to a positive force and then reducing the load to zero and repeating this process until the sample fails with the number of cycles till failure as the desired value to be measured. Tensile creep is similar to this except that the load is not altered but rather steadily applied until the sample fails.

## Types of Materials

Nearly all materials can be tested in tension in one manner or another, but the more popular materials include metals, plastics, woods, polymers and textiles. The test sample often take the shapes of bars, strings, strands, coupons, dog bones, and dumbbells depending upon the material, the tensile grip, and test performed on the sample. Materials with high compressive strength values have relatively low tensile strength, such as brick and aerospace composites. These are not generally tested in tension as their applications do not normally require them to withstand tensile loads.

## Universal Testing Machine

A Universal Testing Machine (UTM) is used to test both the tensile and compressive strength of materials. Universal Testing Machines are named as such because they can perform many different varieties of tests on an equally diverse range of materials, components, and structures. Most UTM models are modular, and can be adapted to fit the customer's needs.

Universal Testing Machines can accommodate many kinds of materials, ranging from hard samples, such as metals and concrete, to flexible samples, such as rubber and textiles. This diversity makes the Universal Testing Machine equally applicable to virtually any manufacturing industry.

The UTM is a versatile and valuable piece of testing equipment that can evaluate materials properties such as tensile strength, elasticity, compression, yield strength, elastic and plastic deformation, bend compression, and strain hardening. Different models of Universal Testing Machines have different load capacities, some as low as 5 kN and others as high as 2,000 kN.

Tests can also be performed in controlled environmental conditions. This is achieved by placing the Universal Testing Machine into an environmental room or chamber. For example, metals testing can be conducted at extreme temperatures: from -196°C (-321°F) to over 1000°C (1800°F).

## Electromechanical Universal Testing Machines

A typical electromechanical UTM repeats a test under the same controlled conditions on test samples, whether it's pushing, pulling, bending, or squashing the sample; the only variable should be the material. You are looking to determine the material's mechanical properties. Just as the speed of a test should be controlled, so should the shape of the sample. Most tensile specimens are designed to have the centre section narrower than the ends, and thus commonly called "dogbone" specimens.

## Data Acquisition in Modern Universal Testing Machines

Measurements of applied force (load) and deformation of the specimen are generally not made directly on the specimen. Instead, electrical signals are sent to a recorder device, typically a computer running data acquisition and analysis software. Of course, the earlier UTM machines that were analog with a chart recorder have been virtually replaced by digital controls and PC-driven software. The typical load/deformation curves represent two components-forces on the Y-axis versus deformation on the X-axis. By far the most common tests performed on plastics with a UTM are tensile strength and modulus and flexural strength and modulus. All testing is done in accordance with specific ASTM and /or ISO standards. Testing in accordance with OEM standards, particularly in the automotive, aerospace and medical device arenas may also be required.

## Container Compression Test

Compression testing is designed to evaluate how much weight a package can withstand, and is particularly relevant to those in the distribution chain and those producing the packaging raw material. It is typically used for evaluating the strength of tertiary packaging, such as stacked boxes, but is equally applicable to stacks of yoghurt pots, for example, and has recently been found to be very useful for testing how much vertical load glass containers can tolerate. The test involves applying an external load to the packaging to replicate the load that is applied during storage and distribution. A load is applied to the container until failure or up to its nominal load.

This testing is intended to recreate compressive loads that may occur during warehouse storage or vehicle transport.

## Reasons of Compression Testing

It is Important in both package and product design verification to determine their performance under compressive loads.

The data generated should provide enough information to:

- Minimize damage costs and assure you that the products packaging will be adequate to give high levels of protection for worldwide distribution.

- Determine if the packaging being used will withstand being stacked and also to what height it can be stacked without damage.

## Corrugated Box Testing

Corrugated shipping containers are exposed to compression hazards during storage and shipment. Proper compression strength is a key performance factor.

## Factors potentially affecting test results

- Size and construction of the specific shipping container under test;

- Grade and flute structure of corrugated fiberboard;

- moisture content of the corrugated board (based on relative humidity);

- Orientation of the box during the test;

- Inner supports, if used during testing (wood, corrugated board, cushioning);

- Contents (when box is tested with contents);

- Box closure;

- Whether the compression machine has "fixed" or "floating" (swiveled) platens;

- Previous handling or testing of box;

- etc.

## Estimations

Corrugated fiberboard can be evaluated by many material test methods including an Edge Crush Test (ECT). There have been efforts to estimate the peak compression strength of a box (usually empty, regular singelwall slotted containers, top-to-bottom) based on various board properties. Some have involved finite element analysis. One of the commonly referenced empirical estimations was published by McKee in 1963. This used the board ECT, the MD and CD flexural stiffness, the box perimeter, and the box depth. Simplifications have used a formula involving the board ECT, the board thickness, and the box perimeter. Most estimations do not relate well to other box orientations, box styles, or to filled boxes. Physical testing of filled and closed boxes remains necessary.

## Calculating Compression requirement

Fiber Box Association have a method for calculating the required compression losses which includes the following factors:

- Time

- Moisture

- Palletizing type

- Pallet patterns

- Pallet type

- Handling

## Dynamic Compression

Containers can be subjected to compression forces that involve distribution dynamics. For example, a package may be impacted by an object being dropped onto it (vertical load) or impacted by freight sliding into it (horizontal load). Vehicle vibration can involve a stack of containers and create dynamic compression responses. Package testing methods are available to evaluate these compression dynamics.

# Stiffness

Stiffness can be defined as the resistance of an object or a system to a change in length. It is calculated by dividing the applied force by the resulting length change.

Very stiff objects or systems only change length by a small amount when a large force is applied. They are stiff.

Less stiff objects or systems change length substantially when the same amount of force is applied. They are compliant.

It is defined as the property of a material which is rigid and difficult to bend. The example of stiffness is rubber band. If single rubber band is stretch by two fingers the stiffness is less and the flexibility is more. Similarly, if we use the set of rubber band and stretched it by two fingers, the stiffness will be more, rigid and flexibility is less.

The expression of stiffness for an elastic body is as below.

$$k = \frac{F}{\delta}$$

Here, the stiffness is k, applied force is F, and deflection is δ.

The unit of stiffness is Newtons per meter.

Stiffness is applied to tension or compression. The length of a bar with cross-sectional area and tensile force applied is shown below in figure.

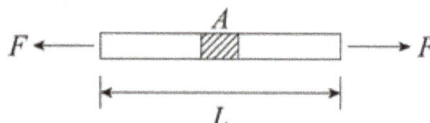

The formula for axial stiffness is expressed as,

$$k = \frac{F}{\dfrac{FL}{AE}}$$

$$k = \frac{FAE}{FL}$$

$$k = \frac{AE}{L}$$

Here, cross-sectional area of an object is A, length is L, applied force is F, and elastic modulus is E.

Stiffness generally depends on the modulus of elasticity, and strength depends on the material's yield strength or its equivalent. For a beam (bending member) stiffness also depends on its moment of Inertia (I) and length. To increase stiffness (to reduce deformation under load) it is most effective to make the member deeper (larger h) since increasing depth rapidly increases the moment of inertia ($I = \dfrac{bh^3}{12}$ for a rectangular shape). For a rectangular beam, b is the width, and h represents the depth.

## Bending Stiffness

Bending stiffness is a term used in materials science for a structure's rigidity and elasticity. The initial stiffness is usually determined using proportional loading of the detail with constant eccentricity e= $M_{Ed}/N_{Ed}$= $const$. Sometimes, stiffness for constant axial force is used.

Simplification deals with the centre of compressive stress under the column flanges. The dependency of deformations $\delta_t$, $\delta_c$ of particular components on inner forces is based on stiffness $k_t$ of part in tension and stiffness $k_c$ of part in compression. Following expressions are used:

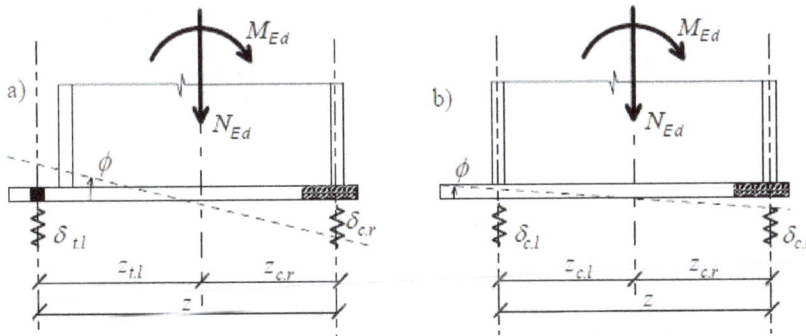

Analysis model

$$\delta_{t,l} = \frac{\dfrac{M_{Ed}}{\approx} - \dfrac{N_{Ed} z_{c,r}}{\approx}}{E k_{t,l}} = \frac{M_{Ed} - N_{Ed} z_{c,r}}{E z k_{t,l}}$$

$$\delta_{t,l} = \frac{\frac{M_{Ed}}{\approx} + \frac{N_{Ed}}{\approx} \approx_{t,l}}{Ek_{c,r}} = \frac{M_{Ed} + N_{Ed}z_{t,l}}{Ezk_{c,r}}$$

The rotation of the foundation is calculated using expression:

$$\phi = \frac{\delta_{t,l} + \delta_{c,r}}{z} = \frac{1}{Ez^2}\left(\frac{M_{Ed} - N_{Ed}z_{c,r}}{k_{t,l}} + \frac{M_{Ed} + N_{Ed}z_{t,l}}{k_{c,r}}\right)$$

Therefore, initial stiffness is

$$S_{j,ini} = \frac{Ez^2}{\frac{1}{k_{c,r}} + \frac{1}{k_{t,l}}} = \frac{Ez^2}{\sum\frac{1}{k}}$$

Non-linear part of moment-rotation dependency is based on the shape factor

$$\mu = \left(1.5\frac{M_{Ed}}{M_{Rd}}\right)^{2.7} \geq 1.0$$

The tangential bending stiffness is calculated using expression

$$S_j = \frac{Ez^2}{\mu\sum\frac{1}{k}}$$

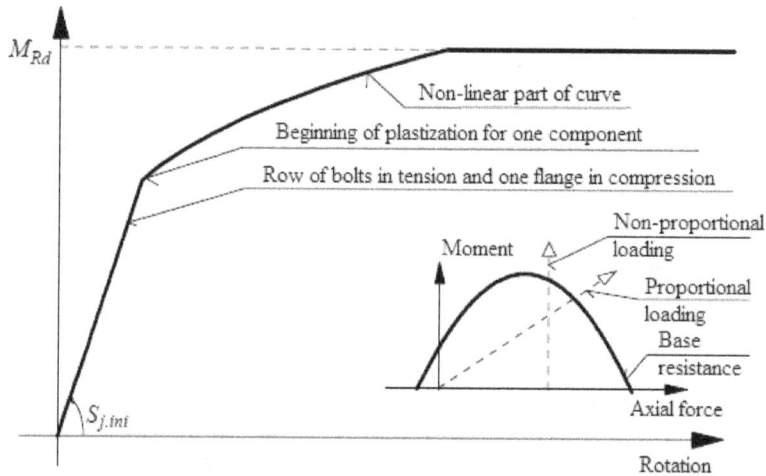

Rotation-moment diagram for proportional loading with constant eccentricity

The linear part of the curve represents bolts in tension and second flange in compression. Non-linear part describes plastic deformation of a component (end plate, bolts in tension or concrete in compression).

Following figure shows the effect of simplified model on resistance determination.

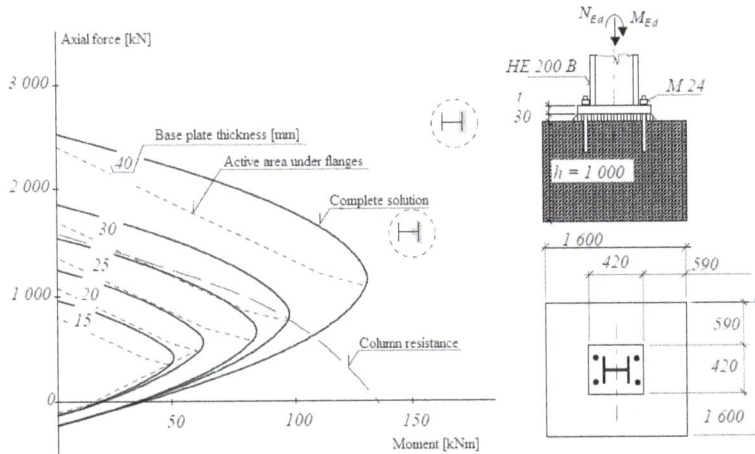

Comparison of simplified model with active area under the column flanges
(curvature changes are caused by the activity of bolts) and model
with active area under the all cross-section

# Deflection

Deflection refers to the movement of a beam or node from its original position due to the forces and loads being applied to the member. Deflection, also known as displacement, can occur from external applied loads or from the weight of the structure itself, and the force of gravity in which this applies. It can occur in beams, trusses, frames and basically any other structure. To define deflection, let's take a simple cantilevered beam deflection that has a person with weight (W) standing at the end:

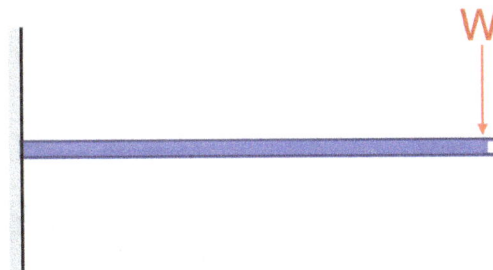

The force of this person standing at the end will cause the beam to bend and deflect from its natural position. In the below diagram the blue beam is the original position, and the dotted line simulates the cantilever beam deflection:

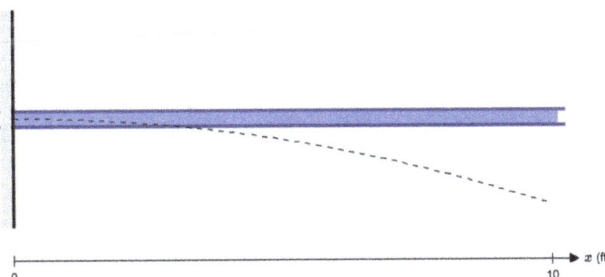

As you can see, the beam has bent or moved away from the original position. This distance at each point along the member is the meaning or definition of deflection.

## Equation of Deflection

Deflection of a beam (beam deflection) is calculated based on a variety of factors, including materials, moment of inertia of a section, the force applied and the distance from a support. There are a range of beam deflection equations that can be used to calculate a basic value for deflection in different types of beams. Generally, deflection can be calculated by taking the double integral of the Bending Moment Equation, M(x) divided by EI (Young's Modulus x Moment of Inertia).

## Unit of Deflection

The unit of deflection, or displacement, is a length unit and normally taken as mm (for metric) and in (for imperial). This number defines the distance in which the beam has deflected from the original position.

## References

- Oral, E; Christensen, SD; Malhi, AS; Wannomae, KK; Muratoglu, OK (2006). "PubMed Central, Table 3:". JArthroplasty. 21:58091. doi:10.1016/j.arth.2005.07.009. PMC 2716092 . PMID 16781413. Archived from the original on 27 April 2018. Retrieved 20 February 2015

- Strength-of-materials: mechanicalc.com, Retrieved 28 June 2018

- "MatWeb – The Online Materials Information Resource". Archived from the original on 21 February 2015. Retrieved 20 February 2015

- Ultimate-tensile-strength-1126: corrosionpedia.com, Retrieved 21 May 2018

- Xu, Wei; Chen, Yun; Zhan, Hang; Nong Wang, Jian (2016). "High-Strength Carbon Nanotube Film from Improving Alignment and Densification". Nano Letters. 16: 946–952. Bibcode:2016NanoL..16..946X. doi:10.1021/acs.nanolett.5b03863

- Compressive-strength: instron.in, Retrieved 26 April 2018

- Barber, A. H.; Lu, D.; Pugno, N. M. (2015). "Extreme strength observed in limpet teeth". Journal of the Royal Society Interface. 12: 105. doi:10.1098/rsif.2014.1326

- Portland-cement-compressive-strength: pavementinteractive.org, Retrieved 06 July 2018

- "Tensile and creep properties of ultra high molecular weight PE fibres" (PDF). Archived from the original (PDF) on 28 June 2007. Retrieved 11 May 2007

- Shear-strength-testing: ametektest.com, Retrieved 15 May 2018

- Yu, Min-Feng; Lourie, O; Dyer, MJ; Moloni, K; Kelly, TF; Ruoff, RS (2000). "Strength and Breaking Mechanism of Multiwalled Carbon Nanotubes Under Tensile Load". Science. 287(5453): 637–640. Bibcode:2000Sci...287..637Y. doi:10.1126/science.287.5453.637. PMID 10649994

- Universal-testing-machines, materials-testing: torontech.com, Retrieved 29 June 2018

- E.J. Pavlina and C.J. Van Tyne, "Correlation of Yield Strength and Tensile Strength with Hardness for Steels", Journal of Materials Engineering and Performance, 17:6 (December 2008)

# Structural Analysis: Tools and Methods

The determination of the influence of loads on structures and their components is known as structural analysis. Computation of a structure's deformations, stresses, stability, internal forces, etc. are also within the scope of this area. This chapter extensively covers the varied tools and methods used in structural analysis, such as direct integration of a beam, Castigliano's method, Macaulay's method, limit state design, moment distribution method, etc.

## Structural Analysis

Structural analysis is an important subject of civil engineering that evaluates the different loads on structures, and their effects. It is an accurate method to ascertain the capability of the structures to withstand the expected loads, and assist in designing the structures accordingly.

Structural analysis is a comprehensive assessment to ensure that the deformations in structural analysis a structure will be adequately lower than the permissible limits, and failure of structural will not occur. The aim of structural analysis is to design a structure that has the proper strength, rigidity, and safety. Deformations in a structure can be either elastic that is totally recoverable, or inelastic that is permanent. Structural analysis assists in the design of structures that meet their functional requirements, are economical and attractive. Structural analysis integrates the disciplines of mechanics, dynamics, and failure theories to compute the internal forces and stresses on the structures to be designed.

Structural analysis is carried out by an examination of the real structure, on a model of the structure created on some scale, and by the utilization of mathematical models. Tests are conducted on the real structure when production is required of similar structures in large quantities, like frames of a particular car, or when the test expenses are acceptable due to the significance of the task. When elements of the main structures are to be examined, then models are used for the estimation of the different loads to be endured. Most structural analyses are conducted on the mathematical models, in which the model could be elastic or inelastic, forces may be static or dynamic, and the model of the structure might be two dimensional or three dimensional.

## Structural Engineering Theory

Structural modeling is an essential and important tool in structural engineering.

Over the past 200 years, many of the most significant contributions to the understanding of the structures have been made by Scientist Engineers while working on mathematical models, which were used for real structures.

Application of mathematical model of any sort to any real structural system must be idealized in some fashion; that is, an analytical model must be developed.

There has never been an analytical model, which is a precise representation of the physical system. While the performance of the structure is the result of natural effects, the development and thus the performance of the model is entirely under the control of the analyst. The validity of the results obtained from applying mathematical theory to the study of the model therefore rests on the accuracy of the model. While this is true, it does not mean that all analytical models must be elaborate, conceptually sophisticated devices. In some cases very simple models give surprisingly accurate results. While in some other cases they may yield answers, which deviate markedly from the true physical behavior of the model, yet be completely satisfactory for the problem at hand.

Structure design is the application of structural theory to ensure that buildings and other structures are built to support all loads and resist all constraining forces that may be reasonably expected to be imposed on them during their expected service life, without hazard to occupants or users and preferably without dangerous deformations, excessive sideways (drift), or annoying vibrations. In addition, good design requires that this objective be achieved economically.

Provision should be made in application of structural theory to design for abnormal as well as normal service conditions. Abnormal conditions may arise as a result of accidents, fire, explosions, tornadoes, severer-than-anticipated earthquakes, floods, and inadvertent or even deliberate overloading of building components. Under such conditions, parts of a building may be damaged. The structural system, however, should be so designed that the damage will be limited in extent and undamaged portions of the building will remain stable. For the purpose, structural elements should be proportioned and arranged to form a stable system under normal service conditions. In addition, the system should have sufficient continuity and ductility, or energy-absorption capacity, so that if any small portion of it should sustain damage, other parts will transfer loads (at least until repairs can be made) to remaining structural components capable of transmitting the loads to the ground.

## Structural Dynamics

Structural dynamics is the study of how structures react to highly varying loads over time. The topic has been historically linked to earthquakes and seismic design of structures and for a long time was considered beyond the daily practice of engineering.

However, this is rapidly changing due to changes in materials, increasingly challenging architectural design, and sustainability considerations. The application of structural dynamics is now highly relevant in different scenarios, such as office floors subject to walking loads and tall buildings subject to wind and seismic loading - as buildings become slimmer and taller, and floors more slender, they exhibit a more significantly dynamic response which affects their functioning.

New bridges have also exhibited a more dynamic response to loads: a famous example is the Millennium Bridge in London. The architectural design dictated that the structure be very slender, which meant that its fundamental natural frequency was low enough to match the exciting frequency of pedestrians using the bridge - resulting in vibrations which caused London's public to name it "the wobbly bridge". After studying the problem, structural engineers added dynamic control devices along the length of the bridge which dampened the vibration.

In very general terms, the research in this area can be split into two categories: the first is research which aims to understand the nature of loads and create loading models. The purpose of these models is to give a mathematical representation which reflects as close as possible these loads, and can help structural engineers produce designs which take account of them. The second area is to define the acceptable performance of structures under these loads and understand how the structures will respond to them.

Trying to understand the nature of the loads requires measuring them in actual structures, then trying to produce a model which represents them as accurately as possible. Technological advances in measurement devices have greatly improved our understanding of such actions, as has the ability to process and analyse huge amounts of data. Advances in mathematics, numerical methods and computer power have contributed hugely to our ability to model and idealize these actions.

Over time the approach to structural dynamics has shifted from designs that aim to make structures *resist* these loads, to make structures *react* to the loads - the application of this is vast, from pedestrian bridges and skyscrapers to a nuclear power plant.

Under this umbrella of "Performance Based Design" is an approach called dynamic control of structures, where devices are added to the structure to influence its behavior under dynamic loading.

Base isolators under the Utah State Capitol

These devices can be either passive or active. An example of a passive device would be base isolation, which is installed at the foundations of structures to absorb seismic energy and prevent it damaging the structure. Active devices are the more sophisticated, reacting to dynamic loading by imposing actions which counteract the external dynamic actions. An example of an active device would be the active mass damper (active control device) used in the Nanjing Communication Tower in China, which helps reduce the effect of lateral loads.

Structural dynamics is proving to be more and more an essential knowledge for structural engineers, as we are challenged by requirements like sustainability and planning restrictions. Better understanding of dynamic behavior will allow us to build longer bridges and taller buildings, while using less material – creating more remarkable structures in a more sustainable way.

# Structural Load

Structural loads or actions are forces, deformations, or accelerations applied to a structure or its components. Assessments of their effects is carried out by the methods of structural analysis. Excess load or overloading may cause structural failure, and such possibility should be either considered in the design or strictly controlled.

The determination of the loads acting on a structure is a complex problem. The nature of the loads varies essentially with the architectural design, the materials, and the location of the structure. Loading conditions on the same structure may change from time to time, or may change rapidly with time.

Loads are usually classified into two broad groups: dead loads and live loads. Dead loads (DL) are essentially constant during the life of the structure and normally consist of the weight of the structural elements. On the other hand, live loads (LL) usually vary greatly. The weight of occupants, snow and vehicles, and the forces induced by wind or earthquakes are examples of live loads. The magnitudes of these loads are not known with great accuracy and the design values must depend on the intended use of the structure.

In structural analysis three kinds of loads are generally used:

1.  Concentrated loads that are single forces acting over a relatively small area, for example vehicle wheel loads, column loads, or the force exerted by a beam on another perpendicular beam.

2.  Line loads that act along a line, for example the weight of a partition resting on a floor, calculated in units of force per unit length.

3.  Distributed (or surface) loads that act over a surface area. Most loads are distributed or are treated as such, for example wind or soil pressure, and the weight of floors and roofing materials.

## Dead Loads

The structure first of all carries the dead load, which includes its own weight, the weight of any permanent non-structural partitions, built-in cupboards, floor surfacing materials and other finishes. It can be worked out precisely from the known weights of the materials and the dimensions on the working drawings. Although the dead load can be accurately determined, it is wise to make a conservative estimate to allow for changes in occupancy; for example, the next owner might wish to demolish some of the fixed partitions and erect others elsewhere.

## Live Loads

All the movable objects in a building such as people, desks, cupboards and filing cabinets produce an imposed load on the structure. This loading may come and go with the result that its intensity will vary considerably. At one moment a room may be empty, yet at another packed with people. Imagine the `extra' live load at a lively party.

# Wind Load

Wind has become a very important load in recent years due to the extensive use of lighter materials and more efficient building techniques. A building built with heavy masonry, timber tiled roof may not be affected by the wind load, but on the other hand the structural design of a modern light gauge steel framed building is dominated by the wind load, which will affect its strength, stability and serviceability. The wind acts both on the main structure and on the individual cladding units. The structure has to be braced to resist the horizontal load and anchored to the ground to prevent the whole building from being blown away, if the dead weight of the building is not sufficient to hold it down. The cladding has to be securely fixed to prevent the wind from ripping it away from the structure.

# Snow Load

The magnitude of the snow load will depend upon the latitude and altitude of the site. In the lower latitudes no snow would be expected while in the high latitudes snow could last for six months or more. In such locations buildings have to be designed to withstand the appropriate amount of snow. The shape of the roof also plays an important part in the magnitude of the snow load. The steeper the pitch, the smaller the load. The snow falling on a flat roof will continue to build up and the load will continue to increase, but on a pitched roof a point is reached when the snow will slide off.

# Earthquake Load

Earthquake loads affect the design of structures in areas of great seismic activity, such as north and South American west coast, New Zealand, Japan, and several Mediterranean countries. Only minor disturbances have been recorded in east Asia and Australia.

# Thermal Loads

All building materials expand or contract with temperature change. Long continuous buildings will expand, and it is necessary to consider the expansion stresses. It is usual to divide a reinforced concrete framed building into lengths not exceeding 30 m and to divide a brick wall into lengths not exceeding 10 m. Expansion joints are provided at these points so that the structure is physically separated and can expand without causing structural damage.

# Settlement Loads

If one part of a building settles more than another part, then stresses are set up in the structures. If the structure is flexible then the stresses will be small, but if the structure is stiff the stresses will be severe unless the two parts of the building are physically separated.

# Dynamic Loads

Dynamic loads, which include impact and aerodynamic loads, are complex. In essence, the magnitude of a load can be greatly increased by its dynamic effect.

## Calculation of Loads

Actual loadings in a building are typically either concentrated or uniformly distributed over an area. The former need no further consideration other than as necessary to characterise them as a force vector. In the latter, however, some modelling is needed when the area considered is actually made up of an assembly of one-way line and surface elements. These elements would pick up different portions of the total load acting over the surface, depending on their arrangement.

Consider the simple structural assembly shown in figure below.

Eight pre-cast concrete elements are supported by three beams Both external beams have to carry the weight of a half concrete element The middle beam carries the weight of one element (½ of the left and right element as illustrated in figure (b)). The reactions from all the elements supported by a beam then become loads acting on the beam. Note that these loads form a continuous line load on the beam. Loads of this type are expressed in terms of a load or force per unit length (i.e. N/m) and are commonly encountered in the structural analysis process.

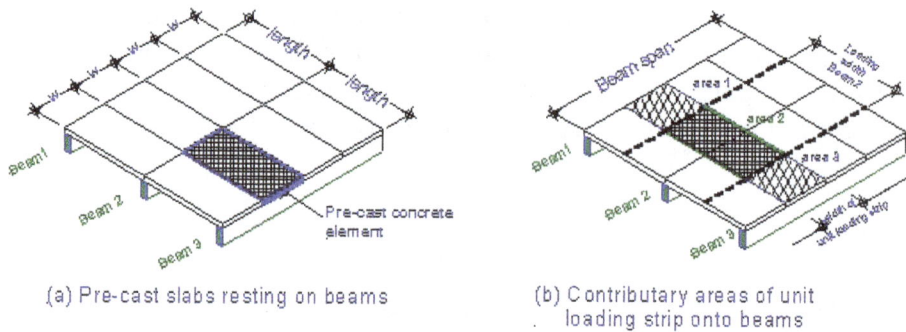

(a) Pre-cast slabs resting on beams

(b) Contributary areas of unit loading strip onto beams

Figure

Another way of looking at this same loading is to think in terms of contributory areas. Each of the beams can be considered as supporting an area of the extent indicated in figure (a) and (b). The width of each area is often called the load strip. The load acting over the width of the load strip is transferred to the support beams. If the uniformly distributed load is constant and the load strip is of a constant width, the amount of load carried per unit length by the support beam is simply the load per unit area multiplied by the width of the load strip. This process is illustrated in figure. The result is again a continuous line load describable in terms of a load per unit length. This process is valid for equal uniformly distributed loads only.

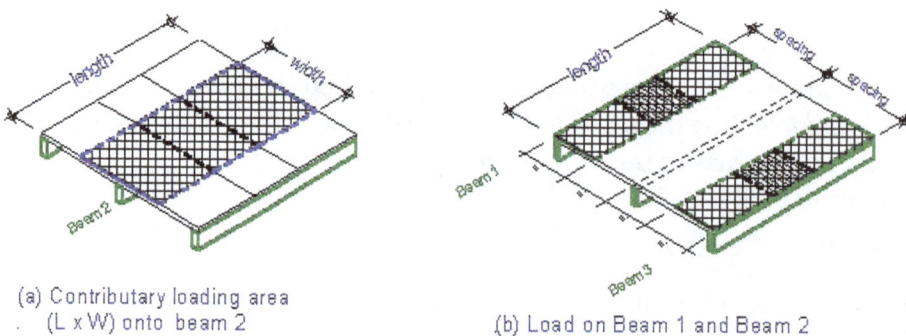

(a) Contributary loading area (L x W) onto beam 2

(b) Load on Beam 1 and Beam 2

Figure

The loading considered should, of course, include both live- and dead-load components. The exact value of the latter can be found by calculating the volume contributary area ´ the thickness of the material and multiply it by the unit weights for that material. Determining these values can be tedious. An alternative is to use a unit weight, e.g. the weight for one square metre, typically expressed as a force per unit area, to represent the weight expressed as N/m², . Since live loads are also expressed in terms of a force per unit area, the calculation process is facilitated, since both loads can be considered simultaneously. Some sample load calculations per m2 are shown below.

## Sample Design Calculations

For design purposes it is most appropriate to select a unit area for all loads (dead, live, wind etc.). This often simplifies the calculation because the unit area may be used for members with the same loading but different contributory areas.

To determine the load per unit area is the most appropriate procedure in structural design. The total load can easily be calculated by load per unit area times the contributory area. For design purposes often the unit loading strip is used as indicated in figure (b) above.

It is convenient to determine first all the loadings per unit area that occur frequently throughout the building. The advantage is that these figures can then be used for all different areas or floor levels with the same loading.

The following is an example of a unit load determination for an office building.

### Flat Roof

| Tanking (Bituminous felt (5-ply) and 50 mm gravel | 1.20 kN/m² |
|---|---|
| 50 mm Insulation | 0.03 " |
| 180 mm Concrete slab (0.18 x 25 kN/m³) | 4.50 " |
| 13 mm Gypsum plaster | 0.22 " |
| DEAD LOAD | 5.95 kN/m² |
| LIVE LOAD (SA 1170.1 & 4.8.1.1) | 0.25 kN/m² |
| TOTAL LOAD | 6.20 kN/m² |

### Offices

| Carpet | 0.05 kN/m² |
|---|---|
| 50 mm Insulation | 0.03 " |
| 200 mm Concrete slab (0.20 x 25 kN/m³) | 5.00 " |
| 13 mm Gypsum plaster | 0.22 " |
| DEAD LOAD | 5.30 kN/m² |
| LIVE LOAD(Appendix B 6.11) | 3.00 kN/m² |
| TOTAL LOAD | 8.30 kN/m² |

## Corridors and Passageways

| Carpet | 0.05 kN/m² |
|---|---|
| 40 mm Screed (0.04 x 22 kN/m³) | 0.88 " |
| 20 mm Insulation | 0.05 " |
| 200 mm Concrete slab (0.20 x 25 kN/m³) | 5.00 " |
| 13 mm Gypsum plaster | 0.22 " |
| DEAD LOAD | 6.20 kN/m² |
| LIVE LOAD(Appendix B 6.4) | 4.00 kN/m² |
| TOTAL LOAD | 10.20 kN/m² |

## Stairs and Landings

| Stairs | |
|---|---|
| Marble tiles | 0.42 kN/m² |
| Concrete wedge (2 x 0.17 x 0.25 x 4 x 23.5 kN/m²) | 2.08 kN/m² |
| 160 mm Concrete slab (0.16 x 25 kN/m²) | 4.00 " |
| DEAD LOAD | 6.50 kN/m² |
| LIVE LOAD(Appendix B 6.4) | 4.00 kN/m² |
| TOTAL LOAD | 10.50 kN/m² |

| Landings | |
|---|---|
| Marble tiles | 0.42 kN/m² |
| 160 mm Concrete slab (0.16 x 25 kN/m²) | 4.00 " |
| DEAD LOAD | 4.42 kN/m² |
| LIVE LOAD(Appendix B 6.4) | 4.00 kN/m² |
| TOTAL LOAD | 8.42 kN/m² |

Having compiled the required unit loading figures the load per running metre for a particular member can be calculated quite quickly by multiplying the unit load with the appropriate depth of the loading strip, or in case the total dead load on a member is needed by multiplying the unit load with the contributory area.

## Moving Load

For beams loaded with concentrated loads, the point of zero shears usually occurs under a concentrated load and so the maximum moment.

Beams and girders such as in a bridge or an overhead crane are subject to moving concentrated loads, which are at fixed distance with each other. The problem here is to determine the moment under each load when each load is in a position to cause a maximum moment. The largest value of these moments governs the design of the beam.

## Single Moving Load

For a single moving load, the maximum moment occurs when the load is at the midspan and the maximum shear occurs when the load is very near the support (usually assumed to lie over the support).

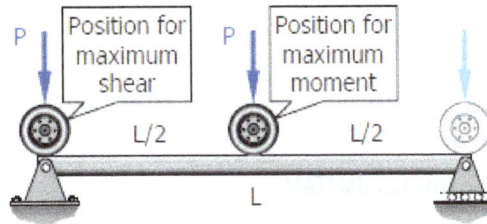

$$M_{max} = \frac{PL}{4} \quad and \quad V_{max} = P$$

## Two Moving Loads

For two moving loads, the maximum shear occurs at the reaction when the larger load is over that support. The maximum moment is given by

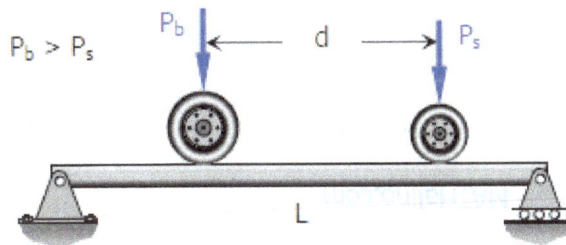

$$M_{max} = \frac{(PL - P_s d)^2}{4PL}$$

where $P_s$ is the smaller load, $P_b$ is the bigger load, and P is the total load ($P = P_s + P_b$).

## Three or More Moving Loads

In general, the bending moment under a particular load is a maximum when the centre of the beam is midway between that load and the resultant of all the loads then on the span. With this rule, we compute the maximum moment under each load, and use the biggest of the moments for the design. Usually, the biggest of these moments occurs under the biggest load.

The maximum shear occurs at the reaction where the resultant load is nearest. Usually, it happens if the biggest load is over that support and as many a possible of the remaining loads are still on the span.

In determining the largest moment and shear, it is sometimes necessary to check the condition when the bigger loads are on the span and the rest of the smaller loads are outside.

Structures that carry moving loads can have finite dimensions or can be infinite and supported periodically or placed on the elastic foundation.

Let us consider simply supported string of the length $l$, cross-sectional area $A$, mass density $\rho$, tensile force $N$, subjected to a constant force $P$ moving with constant velocity $v$. The motion equation of the string under the moving force has a form

$$-N\frac{\partial^2 w(x,t)}{\partial x^2} + \rho A\frac{\partial^2 w(x,t)}{\partial t^2} = \delta(x - vt)P .$$

Displacements of any point of the simply supported string is given by the sinus series

$$w(x,t) = \frac{2P}{\rho A l}\sum_{j=1}^{\infty}\frac{1}{\omega_{(j)}^2 - \omega^2}\left[\sin(\omega t) - \frac{\omega}{\omega_{(j)}}\sin(\omega_{(j)}t)\right],$$

where,

$$\omega = \frac{j\pi v}{l},$$

and the natural circular frequency of the string

$$\omega_{(j)}^2 = \frac{j^2\pi^2}{l^2}\frac{N}{\rho A}.$$

In the case of inertial moving load the analytical solutions are unknown. The equation of motion is increased by the term related to the inertia of the moving load. A concentrated mass $m$ accompanied by a point force $P$:

$$-N\frac{\partial^2 w(x,t)}{\partial x^2} + \rho A\frac{\partial^2 w(x,t)}{\partial t^2} = \delta(x - vt)P - \delta(x - vt)m\frac{\mathrm{d}^2 w(vt,t)}{\mathrm{d}t^2} .$$

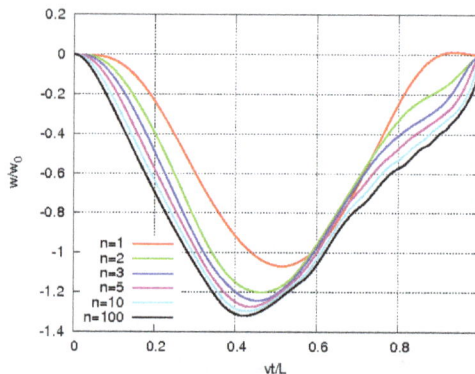

Convergence of the solution for different number of terms.

The last term, because of complexity of computations, is often neglected by engineers. The load influence is reduced to the massless load term. Sometimes the oscillator is placed in the contact point. Such approaches are acceptable only in low range of the travelling load velocity. In higher ranges both the amplitude and the frequency of vibrations differ significantly in the case of both types of a load.

The differential equation can be solved in a semi-analytical way only for simple problems. The series determining the solution converges well and 2-3 terms are sufficient in practice. More complex problems can be solved by the finite element method or space-time finite element method.

**Massless Load**

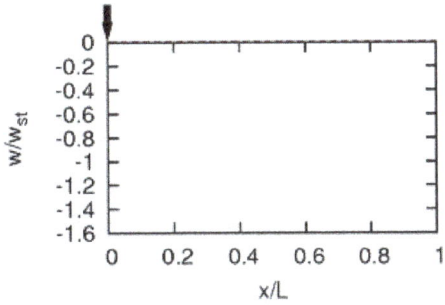

Vibrations of a string under a moving massless
force ($v=0.1c$); $c$ is the wave speed.

**Inertial Load**

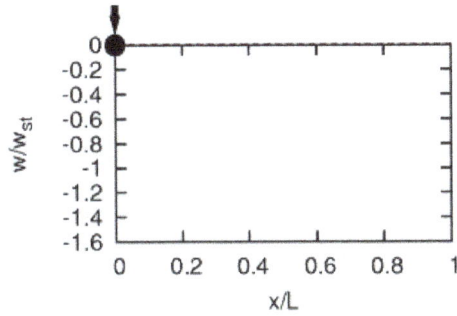

Vibrations of a string under a moving inertial force
($v=0.1c$); $c$ is the wave speed.

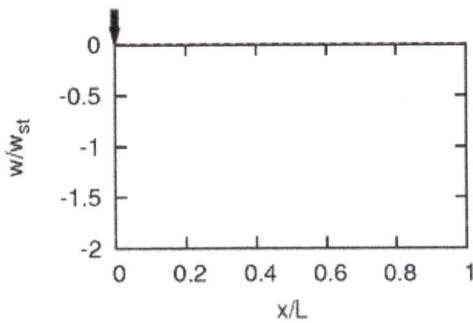

Vibrations of a string under a moving massless
force ($v=0.5c$); $c$ is the wave speed.

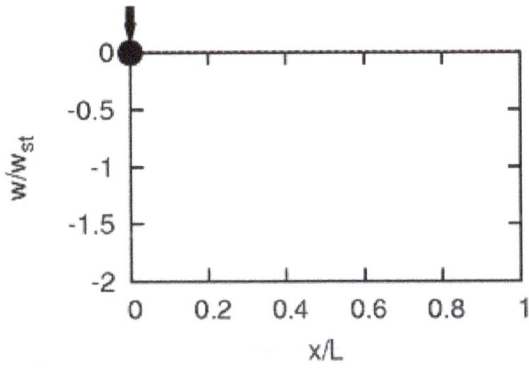

Vibrations of a string under a moving inertial force
($v=0.5c$); $c$ is the wave speed.

The discontinuity of the mass trajectory is also well visible in the Timoshenko beam. High shear
stiffness emphasizes the phenomenon.

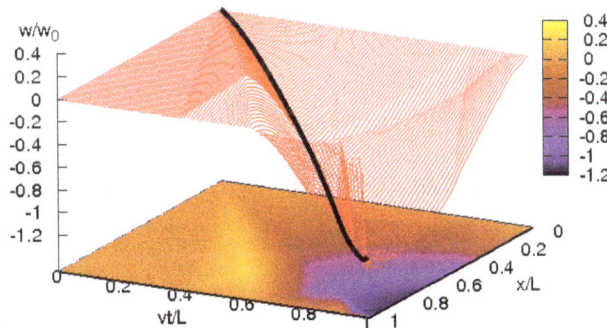

Vibrations of the Timoshenko beam: red lines - beam axes in time,
black line - mass trajectory ($w_o$ - static deflection).

## The Renaudot Approach vs. The Yakushev Approach

The Renaudot approach

$$\delta(x - vt)\frac{d}{dt}\left[m\frac{dw(vt,t)}{dt}\right] = \delta(x - vt)m\frac{d^2w(vt,t)}{dt^2}.$$

The Yakushev approach

$$\frac{d}{dt}\left[\delta(x-vt)m\frac{dw(vt,t)}{dt}\right] = -\delta'(x-vt)mv\frac{dw(vt,t)}{dt} + \delta(x-vt)m\frac{d^2w(vt,t)}{dt^2}.$$

## Massless String Under Moving Inertial Load

Let us consider a massless string, which is a particular case of moving inertial load problem. The first solve the problem Smith. The analysis will follow the solution of Fryba. Assuming ρ=0, the equation of motion of a string under a moving mass can be put into the following form:

$$-N\frac{\partial^2 w(x,t)}{\partial x^2} = \delta(x-vt)P - \delta(x-vt)m\frac{d^2w(vt,t)}{dt^2}.$$

We impose simply-supported boundary conditions and zero initial conditions. To solve this equation we use the convolution property. We assume dimensionless displacements of the string y and dimensionless time τ:

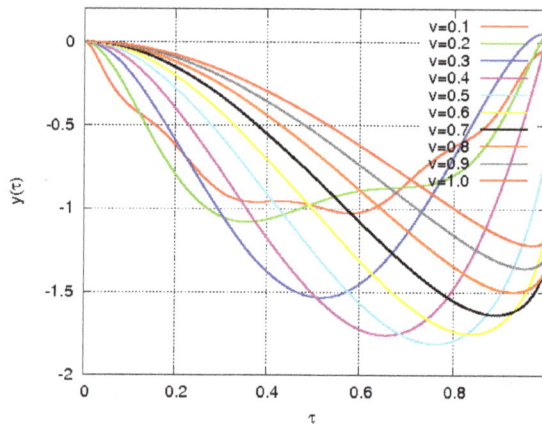

Massless string and a moving mass - mass trajectory.

$$y(\tau) = \frac{w(vt,t)}{w_{st}}, \quad \tau = \frac{vt}{l},$$

where $w_{st}$ is the static deflection in the middle of the string. The solution is given by a sum

$$y(\tau) = \frac{4\alpha}{\alpha-1}\tau(\tau-1)\sum_{k=1}^{\infty}\prod_{i=1}^{k}\frac{(a+i-1)(b+i-1)}{c+i-1}\frac{\tau^k}{k!},$$

where α is the dimensionless parameters :

$$\alpha = \frac{Nl}{2mv2} > 0 \quad \wedge \quad \alpha \neq 1.$$

Parameters a, b and c are given below

$$a_{1,2} = \frac{3\pm\sqrt{1+8\alpha}}{2}, \quad b_{1,2} = \frac{3\mp\sqrt{1+8\alpha}}{2}, \quad c = 2.$$

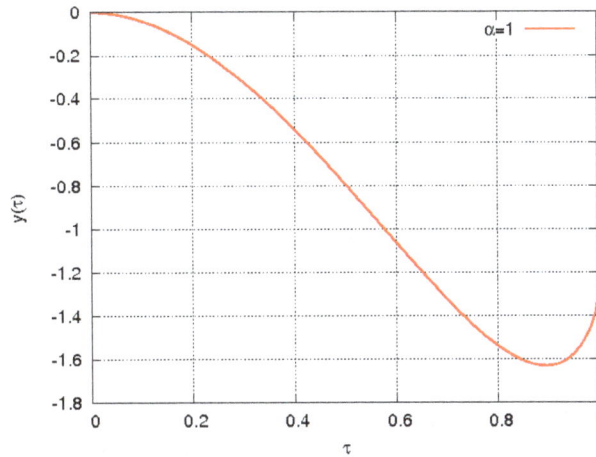

Massless string and a moving mass - mass trajectory, α=1.

In the case of α=1 the considered problem has a closed solution

$$y(\tau) = \left[\frac{4}{3}\tau(1-\tau) - \frac{4}{3}\tau\left(1 + 2\tau \ln(1-\tau) + 2\ln(1-\tau)\right)\right].$$

## Direct Integration of a Beam

The deflection of the loaded beam can be obtained by various methods,

Direct integration method: The governing differential equation is defined as,

$$M = EI\frac{d^2 y}{dx^2} \quad \text{or} \quad \frac{M}{EI} = \frac{d^2 y}{dx^2}$$

on integrating one get,

$$\frac{dy}{dx} = \int \frac{M}{EI} dx + A \ldots \text{ this equation gives the slope}$$

of the loaded beam

Integrate once again to get the deflection.

$$y = \int\int \frac{M}{EI} dx + Ax + B$$

Where A and B are constants of integration to be evaluated from the known conditions of slope and deflections for the particular value of x.

Illustrative examples : let us consider few illustrative examples to have a familiarty with the direct integration method.

Case 1: Cantilever Beam with Concentrated Load at the end:- A cantilever beam is subjected to a concentrated load W at the free end, it is required to determine the deflection of the beam.

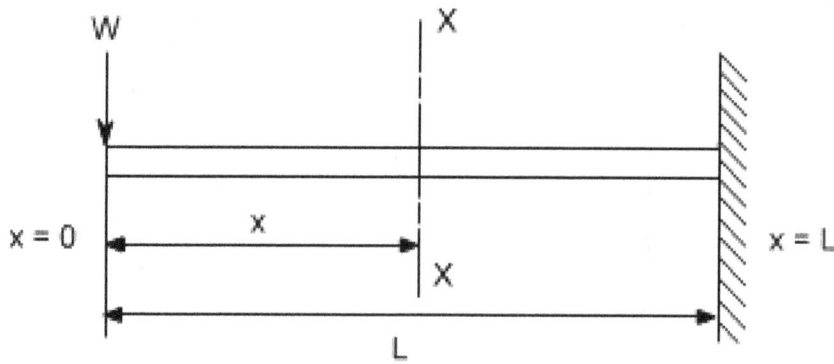

In order to solve this problem, consider any X-section X-X located at a distance x from the left end or the reference, and write down the expressions for the shear force abd the bending moment.

$$S.F\big|_{x\text{-}x} = -W$$

$$B.M\big|_{x\text{-}x} = -W.x$$

$$\text{Therefore } M\big|_{x\text{-}x} = -W.x$$

the governing equation $\dfrac{M}{EI} = \dfrac{d^2 y}{dx^2}$

substituting the value of M interms of x then integrating the equation one get ,

$$\frac{M}{EI} = \frac{d^2 y}{dx^2}$$

$$\frac{d^2 y}{dx^2} = -\frac{Wx}{EI}$$

$$\int \frac{d^2 y}{dx^2} = \int -\frac{Wx}{EI} dx$$

$$\frac{dy}{dx} = -\frac{Wx^2}{2EI} + A$$

Integrating once more,

$$\int \frac{dy}{dx} = \int -\frac{Wx^2}{2EI} dx + \int A dx$$

$$y = -\frac{Wx^3}{6EI} + Ax + B$$

The constants A and B are required to be found out by utilizing the boundary conditions as defined below:

i.e at x= L ; y= o

at x = L ; dy/dx = o

Utilizing the second condition, the value of constant A is obtained as

$$A = \frac{W I^2}{2\,EI}$$

While employing the first condition yields

$$y = -\frac{WL^3}{6\,EI} + AL + B$$

$$B = \frac{WL^3}{6\,EI} - AL$$

$$= \frac{WL^3}{6\,EI} - \frac{WL^3}{2\,EI}$$

$$= \frac{WL^3 - 3\,WL^3}{6\,EI} = -\frac{2\,WL^3}{6\,EI}$$

$$B = -\frac{WL^3}{3\,EI}$$

Subsituting the values of A and B we get

$$y = \frac{1}{EI}\left[ -\frac{W x^3}{6\,EI} + \frac{WL^2 x}{2\,EI} - \frac{WL^3}{3\,EI} \right]$$

The slope as well as the deflection would be maximum at the free end hence putting x = 0 we get,

$$y_{max} = -\frac{WL^3}{3\,EI}$$

$$\left( slope \right)_{max} m = +\frac{WL^2}{2\,EI}$$

Case 2: A Cantilever with Uniformly distributed Loads:- In this case the cantilever beam is subjected to U.d.l with rate of intensity varying w / length. The same procedure can also be adopted in this case.

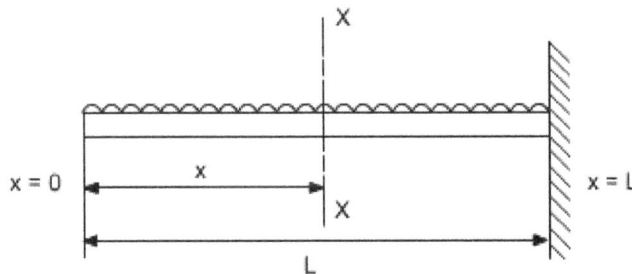

$$S.F\big|_{x\text{-}x} = -w$$

$$B.M\big|_{x\text{-}x} = -w.x.\frac{x}{2} = w\left( \frac{x^2}{2} \right)$$

$$\frac{M}{EI} = \frac{d^2 y}{d x^2}$$

$$\frac{d^2 y}{dx^2} = -\frac{wx^2}{2EI}$$

$$\int \frac{d^2 y}{dx^2} = \int -\frac{wx^2}{2EI} dx$$

$$\frac{dy}{dx} = -\frac{wx^3}{6EI} + A$$

$$\int \frac{dy}{dx} = \int -\frac{wx^3}{6EI} dx + \int A dx$$

$$y = -\frac{wx^4}{6EI} + Ax + B$$

Boundary conditions relevant to the problem are as follows:

1.  At x = L; y = o

2.  At x= L; dy/dx = o

The second boundary conditions yields

$$A = +\frac{wx^3}{6EI}$$

whereas the first boundary conditions yields

$$B = \frac{wL^4}{24EI} - \frac{wL^4}{6EI}$$

$$B = -\frac{wL^4}{8EI}$$

Thus, $y = \frac{1}{EI}\left[ -\frac{wx^4}{24} + \frac{wL^3 x}{6} - \frac{wL^4}{8} \right]$

So $y_{max^m}$ will be at x = o

$$y_{max^m} = -\frac{wL^4}{8EI}$$

$$\left(\frac{dy}{dx}\right)_{max^m} = \frac{wL^3}{6EI}$$

Case 3: Simply Supported beam with uniformly distributed Loads:- In this case a simply supported beam is subjected to a uniformly distributed load whose rate of intensity varies as w / length.

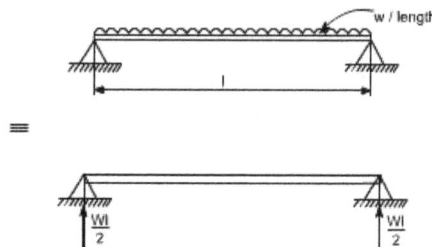

In order to write down the expression for bending moment consider any cross-section at distance of x metre from left end support.

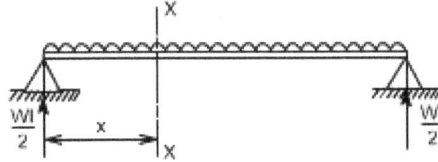

$$S.F\big|_{x\text{-}x} = w\left(\frac{1}{2}\right) - w.x$$

$$B.M\big|_{x\text{-}x} = w.\left(\frac{1}{2}\right).x - w.x.\left(\frac{x}{2}\right)$$

$$= \frac{wl.x}{2} - \frac{wx^2}{2}$$

The differential equation which gives the elastic curve for the deflected beam is ;

$$\frac{d^2 y}{dx^2} = \frac{M}{EI} = \frac{1}{EI}\left[\frac{wl.x}{2} - \frac{wx^2}{2}\right]$$

$$\frac{dy}{dx} = \int \frac{wlx}{2EI}dx - \int \frac{wx^2}{2EI}dx + A$$

$$= \frac{wlx^2}{4EI} - \frac{wx^3}{6EI} + A$$

Integrating, once more one gets

$$y = \frac{wlx^3}{12EI} - \frac{wx^4}{24EI} + A.x + B$$

Boundary conditions which are relevant in this case are that the deflection at each support must be zero.

i.e. at x = 0; y = 0 : at x = l; y = 0

let us apply these two boundary conditions on equation ( $y = \frac{wlx^3}{12EI} - \frac{wx^4}{24EI} + A.x + B$ ) because the boundary conditions are on y, This yields B = 0.

Futher

$$0 = \frac{wl^4}{12EI} - \frac{wl^4}{24EI} + A.l$$

$$A = -\frac{wl^3}{24EI}$$

So the equation which gives the deflection curve is ;

$$y = \frac{1}{EI}\left[\frac{wLx^4}{12} - \frac{wx^4}{24} - \frac{wL^3x}{24}\right]$$

In this case the maximum deflection will occur at the centre of the beam where x = L/2 [ i.e. at the position where the load is being applied ].So if we substitute the value of x = L/2

Then $\quad y_{max^m} = \dfrac{1}{EI}\left[\dfrac{wL}{12}\left(\dfrac{L^3}{8}\right) - \dfrac{w}{24}\left(\dfrac{L^4}{16}\right) - \dfrac{wL^3}{24}\left(\dfrac{L}{2}\right)\right]$

$$y_{max^m} = -\dfrac{5wL^4}{384EI}$$

Conclusions:

(i)  The value of the slope at the position where the deflection is maximum would be zero.

(ii) The value of maximum deflection would be at the center i.e. at x = L/2.

The final equation which is governs the deflection of the loaded beam in this case is:

$$y = \dfrac{1}{EI}\left[\dfrac{wLx^3}{12} - \dfrac{wx^4}{24} - \dfrac{wL^3x}{24}\right]$$

By successive differentiation one can find the relations for slope, bending moment, shear force and rate of loading.

## Deflection (y)

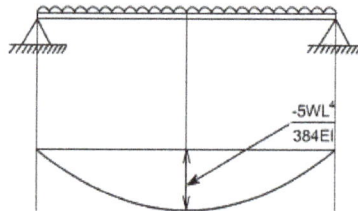

$$y = \dfrac{1}{EI}\left[\dfrac{wLx^3}{12} - \dfrac{wx^4}{24} - \dfrac{wL^3x}{24}\right].$$

## Slope (dy/dx)

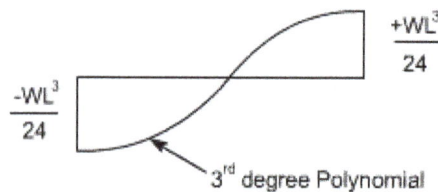

So the bending moment diagram would be:

$$EI.\dfrac{dy}{dx} = \left[\dfrac{3wLx^2}{12} - \dfrac{4wx^3}{24} - \dfrac{wL^3}{24}\right]$$

## Bending Moment

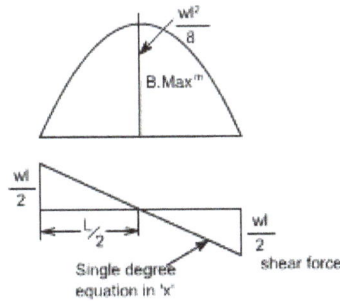

$$\frac{d^2 y}{dx^2} = \frac{1}{EI}\left[\frac{wLx}{2} - \frac{wx^2}{2}\right]$$

## Shear Force

Shear force is obtained by taking

third derivative.

$$EI\frac{d^3 y}{dx^3} = \frac{wL}{2} - w.x$$

## Rate of Intensity of Loading

$$EI\frac{d^4 y}{dx^4} = -w$$

Case 4: The direct integration method may become more involved if the expression for entire beam is not valid for the entire beam. Let us consider a deflection of a simply supported beam which is subjected to a concentrated load W acting at a distance 'a' from the left end.

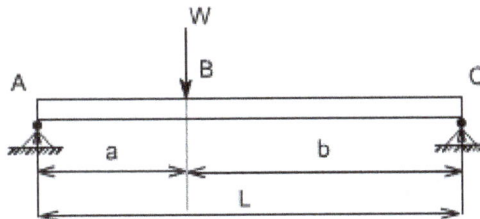

Let $R_1$ & $R_2$ be the reactions then,

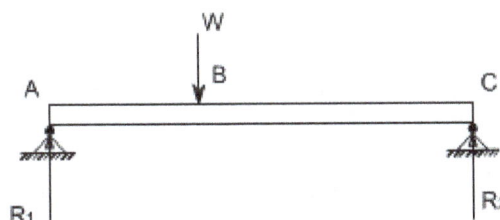

B.M for the portion AB

$M\big|_{AB} = R_1 . x \quad 0 \leq x \leq a$

B.M for the portion BC

$M\big|_{BC} = R_1 . x - W(x-a) \quad a \leq x \leq \big|$

so the differential equation for the two cases would be,

$$EI \frac{d^2 y}{dx^2} = R_1 x$$

$$EI \frac{d^2 y}{dx^2} = R_1 x - W(x-a)$$

These two equations can be integrated in the usual way to find 'y' but this will result in four constants of integration two for each equation. To evaluate the four constants of integration, four independent boundary conditions will be needed since the deflection of each support must be zero, hence the boundary conditions (a) and (b) can be realized.

Further, since the deflection curve is smooth, the deflection equations for the same slope and deflection at the point of application of load i.e. at x = a. Therefore four conditions required to evaluate these constants may be defined as follows:

(a) at x = 0; y = 0 in the portion AB i.e. $0 \leq x \leq a$

(b) at x = 1; y = 0 in the portion BC i.e. $a \leq x \leq 1$

(c) at x = a; dy/dx, the slope is same for both portion

(d) at x = a; y, the deflection is same for both portion

By symmetry, the reaction $R_1$ is obtained as

$$R_1 = \frac{Wb}{a+b}$$

Hence,

$$EI \frac{d^2 y}{dx^2} = \frac{Wb}{(a+b)} x \qquad 0 \leq x \leq a$$

$$EI \frac{d^2 y}{dx^2} = \frac{Wb}{(a+b)} x - W(x-a) \quad a \leq x \leq 1$$

integrating (1) and (2) we get,

$$EI \frac{dy}{dx} = \frac{Wb}{2(a+b)} x^2 + k_1 \qquad 0 \leq x \leq a$$

$$EI \frac{dy}{dx} = \frac{Wb}{2(a+b)} x^2 - \frac{W(x-a)^2}{2} + k_2 \quad a \leq x \leq 1$$

Using condition (c) in equation ($EI \frac{dy}{dx} = \frac{Wb}{2(a+b)} x^2 + k_1$) and ($EI \frac{dy}{dx} = \frac{Wb}{2(a+b)} x^2 - \frac{W(x-a)^2}{2} + k_2$) shows that these constants should be equal, hence letting,

$K_1 = K_2 = K$

Hence,

$$El\frac{dy}{dx} = \frac{Wb}{2(a+b)}x^2 + k \quad o \le x \le a$$

$$El\frac{dy}{dx} = \frac{Wb}{2(a+b)}x^2 - \frac{W(x-a)^2}{2} + k \ a \le x \le l$$

Integrating agian equation $\left(El\frac{dy}{dx} = \frac{Wb}{2(a+b)}x^2\right)$ and $\left(El\frac{dy}{dx} = \frac{Wb}{2(a+b)}x^2 - \frac{W(x-a)^2}{2} + k \ a \le x \le l\right)$ we get

$$Ely = \frac{Wb}{6(a+b)}x^3 + kx + k_3 \qquad o \le x \le a$$

$$Ely = \frac{Wb}{6(a+b)}x^3 - \frac{W(x-a)^3}{6} + kx + k_4 \qquad a \le x \le l$$

Utilizing condition (a) in equation $\left(Ely = \frac{Wb}{6(a+b)}x^3 + kx + k_3\right)$ yields

$$k_3 = o$$

Utilizing condition (b) in equation $\left(Ely = \frac{Wb}{6(a+b)}x^3 - \frac{W(x-a)^3}{6} + kx + k_4\right)$ yields

$$0 = \frac{Wb}{6(a+b)}l^3 - \frac{W(l-a)^3}{6} + kl + k_4$$

$$k_4 = -\frac{Wb}{6(a+b)}l^3 + \frac{W(l-a)^3}{6} - kl$$

But $a + b = l$,

Thus,

$$k_4 = -\frac{Wb(a+b)^2}{6} + \frac{Wb^3}{6} - k(a+b)$$

Now lastly $k_3$ is found out using condition (d) in equation ($Ely = \frac{Wb}{6(a+b)}x^3 + kx + k_3$) and equation ($Ely = \frac{Wb}{6(a+b)}x^3 - \frac{W(x-a)^3}{6} + kx + k_4$), the condition (d) is that,

At $x = a$; $y$; the deflection is the same for both portion.

Therefore $y\Big|_{\text{from equation 5}} = y\Big|_{\text{from equation 6}}$

$$\frac{Wb}{6(a+b)}x^3 + kx + k_3 = \frac{Wb}{6(a+b)}x^3 - \frac{W(x-a)^3}{6} + kx + k_4$$

$$\frac{Wb}{6(a+b)}a^3 + ka + k_3 = \frac{Wb}{6(a+b)}a^3 - \frac{W(a-a)^3}{6} + ka + k_4$$

Thus, $k_4 = o$;

OR

$$k_4 = -\frac{Wb(a+b)^2}{6} + \frac{Wb^3}{6} - k(a+b) = 0$$

$$k(a+b) = -\frac{Wb(a+b)^2}{6} + \frac{Wb^3}{6}$$

$$k = \frac{Wb(a+b)}{6} + \frac{Wb^3}{6(a+b)}$$

so the deflection equations for each portion of the beam are

$$Ely = \frac{Wb}{6(a+b)}x^3 + kx + k_3$$

$$Ely = \frac{Wbx^3}{6(a+b)} - \frac{Wb(a+b)x}{6} + \frac{Wb^3 x}{6(a+b)} \qquad .... \text{for } o \pounds x \pounds a$$

and for other portion

$$Ely = \frac{Wb}{6(a+b)}x^3 - \frac{W(x-a)^3}{6} + kx + k_4$$

Substituting the value of 'k' in the above equation

$$Ely = \frac{Wbx^3}{6(a+b)} - \frac{W(x-a)^3}{6} - \frac{Wb(a+b)x}{6} + \frac{Wb^3 x}{6(a+b)} \qquad \text{For for } a \le x \le l$$

so either of the equation $\left( Ely = \frac{Wbx^3}{6(a+b)} - \frac{Wb(a+b)x}{6} + \frac{Wb^3 x}{6(a+b)} \right)$ or

$\left( Ely = \frac{Wbx^3}{6(a+b)} - \frac{W(x-a)^3}{6} - \frac{Wb(a+b)x}{6} + \frac{Wb^3 x}{6(a+b)} \right)$ may be used to find the deflection at $x = a$

hence sub stituting $x = a$ in either of the equation we get

$$y\big|_{x=a} = -\frac{Wb a^2 b^2}{3El(a+b)}$$

OR if $a = b = 1/2$

$$y_{maxm} = -\frac{WL^3}{48El}$$

Alternate Method: There is also an alternative way to attempt this problem in a more simpler way. Let us considering the origin at the point of application of the load,

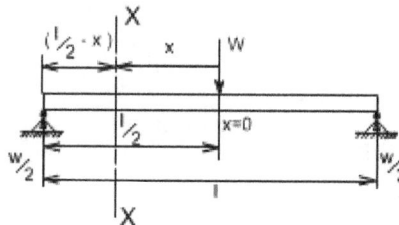

$$S.F\big|_{xx} = \frac{W}{2}$$

$$B.M\big|_{xx} = \frac{W}{2}\left(\frac{1}{2} - x\right)$$

substituting the value of Min the governing equation for the deflection :

$$\frac{d^2y}{dx^2} = \frac{\dfrac{W}{2}\left(\dfrac{1}{2}-x\right)}{El}$$

$$\frac{dy}{dx} = \frac{1}{El}\left[\frac{WLx}{4} - \frac{Wx^2}{4}\right] + A$$

$$y = \frac{1}{El}\left[\frac{WLx^2}{8} - \frac{Wx^2}{12}\right] + Ax + B$$

*Boundary conditions relevant for this case are as follows :*

(*i*) *at x* = 0; *dy / dx* = 0

*hence, A* = 0

(*ii*) *at x* = *l/2; y* = 0 (*because now l / 2 is on the left end or right end support since we have taken the origin at the centre*)

Thus,

$$0 = \left[\frac{WL^3}{32} - \frac{WL^3}{96} + B\right]$$

$$B = -\frac{WL^3}{48}$$

Hence he equation which governs the deflection would be,

$$y = \frac{1}{El}\left[\frac{WLx^2}{8} - \frac{Wx^2}{12} - \frac{WL^3}{48}\right]$$

Hence

$$y_{max}{}^m \big|_{at\,x=0} = -\frac{WL^3}{48\,El} \qquad \text{At the centre}$$

$$\left(\frac{dy}{dx}\right)_{max}{}^m \bigg|_{at\,x=\pm\frac{L}{2}} = \pm\frac{WL^3}{16\,El} \quad \text{At the ends}$$

Hence the integration method may be bit cumbersome in some of the case. Another limitation of the method would be that if the beam is of non uniform cross section,

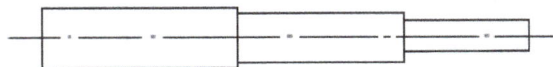

i.e. it is having different cross-section then this method also fails.

## Castigliano's Method

This method deals with displacements with respect to strain energy (Strain energy is also called potential energy, also called Internal energy). It is really the strain energy method of finding deflection. Strain energy is recoverable as mechanical energy or mechanical work.

## Displacement in Castigliano's Method Means

- Extension under an axial tensile force, $\delta L = PL/AE$.

- Contraction under an axial compressive force, $\delta L = -PL/AE$.

- Deflection due to Bending Moment and shear force, $y = \delta_M + \delta_V$.

- Angle of twist due torque (torsional moment), $\Theta = TL/GJ$.

- Extension in a spring due to a force, $\delta = W/k$.

- Contraction in a spring due to a force, $\delta = W/k$ and etc.

## Castigliano's Method

- Castigliano's Method determines the displacement in a linearly elastic system (loaded bar, beam, shaft or structure) at any point by using the partial derivatives of the strain energy. There are two Castigliano's theorems.

(i) Castigliano's first theorem or castigliano's displacement finding method or castigliano's displacement theorem,

It determines Displacement at a point of the loaded beam by using the partial derivatives of the strain energy with respect to the force acting at that location. If the Displacement is positive, then it is in the direction of the force and vice versa.

In the mathematical form, it is given as;

$$\delta_i = \partial U / \partial F_i$$

Where,

$\delta_i$ is the displacement of the point under the force $F_i$ in the direction of $F_i$

U is the total strain energy,

COR: Castigliano's theorem can also determine the slope at a point of the loaded beam by using the partial derivatives of the strain energy with respect to the bending moment acting at that location.

In the mathematical form, it is given as:

$$\theta_i = \partial U / \partial M_i$$

Where U is the total strain energy, $\theta_i$ is the slope (rotational displacement in radians) of the moment $M_i$ in the direction of M (Clockwise or anticlockwise).

note: this method is very useful for obtaining the deflection of a point where there is no force or no moment. in this case a force/ moment (fictitious) is assumed to act and then made zero at that point in the analysis after finding the derivative of the strain energy with respect to the force.

(ii) castigliano's second theorem or castigliano's force finding method or castigliano's force theorem.

It determines the force at a certain point of known displacement '$\delta_i$' within elastic limits. Express strain energy of an elastic structure as a function of generalized displacement $\delta_i$; then find the partial derivative of the strain energy with respect to generalized displacement, $\delta_i$. This partial derivative gives the force $F_i$.

In the mathematical form, it is given as;

$$F_i = \partial U / \partial \delta_i$$

Where U is the total strain energy due to the forces/moments/torques etc.

$\delta_i$ is the displacement at any point 'i'

$F_i$ is the force at the point 'i'

## Meaning of Generalized Displacements

1. In Simple Tension

    Extension in tension $\delta L = PL / AE$

    Strain energy = Average Force $\times \delta L = (F / 2)(PL / AE) = = F^2 L / 2EA$

2. In Simple Compression

    Extension in compression $\delta L = PL / AE$

    Strain energy $= F_{av} \delta L = (F / 2)(PL / AE) = = F^2 L / 2EA$

3. In Simple Bending, Strain energy = Average Bending moment x deflection

    $= (M / 2)(ML / EI) = M^2 L / 2EI$

4. Angle of twist in simple torsion,

    Strain energy= Average torque x displacement =average torque x angle of twist

    $= (T / 2)(TL / GJ) = T^2 L / 2GJ$

## Strain Energy in a Member for any One Type of Loading

| Sr. No. | Type of loading | Strain energy with variable loading | Strain energy with constant loading |
|---|---|---|---|
| 1. | Axial load(tensile or compressive), F | $U = \int F^2 dx / 2EA$ | $U = F^2 L / 2EA$ |
| 2. | Bending Moment, M | $U = \int M^2 dx / 2EA$ | $U = M^2 L / 2EI$ |

| 3. | Torque T | $U = \int T^2 dx / 2GJ$ | $U = T^2 L / 2GJ$ |
|---|---|---|---|
| 4. | Direct shear force, F | $U = \int F^2 dx / 2GA$ | $U = F^2 L / 2GA$ |
| 5. | Transverse shear force, V | $U = \int V^2 dx / 2GA$ | $U = K V^2 L / 2GA$ |

Integration Limits are from 0 to L in the tabular equations.

Where K is a correction factor depending upon the shape of the area

K = 1.2 for a rectangular section

K = 1.11 for a solid circular section

K = 2.0 for a thin walled tube

## Procedure for Application of Castigliano's Theorem

a) To determine a deflection $y_f$ in the direction of a real force $F_f$ or fictitious force $F_f$

1. Obtain an expression for the total strain energy.

2. Obtain the linear deflection $y_f$ from the relationship $y_f = \partial U / \partial F_f$.

3. If the force is fictitious, set $F_f = 0$ and solve the resulting equationa).

4. To determine the slope (an angular displacement) $\theta_f$ in the direction of a fictitious moment $M_f$.

5. Obtain an expression for the total strain energy.

6. Obtain the angular deflection from the relationship $\theta_f = \partial U / \partial M_f$.

7. If the moment is fictitious, then set $M_f = 0$ and solve the resulting equation.

## Example on Castigliano's Theorem

For a given rectangular beam under a central load W, find the deflection under the load.

*Given* : $L = 2m$, $b = 0.1m$, $h = 0.05m$, $W = 40\ 000\ N$, $E = 200\ GPa$, $G = 80\ GPa$ and $I = 417 \times 10^{-6} m^4$

For this example, the beam is of rectangular section, width b and depth h. Total strain energy will be due to bending moment and due to traverse shear. Because the beam is symmetrical, Calculation is made for L/2 and then made 2 times for the entire beam.

Consider a point at distance x from the left hand support.

Moment M = (W/2) x and Transverse Shear Force V = W/2

1. The expression for the total strain energy in length L = 2 times that in length L/2.

$$U = \int M^2 dx / 2EI + 2V^2 dx / 2GA \; (M = W\,x/2 \;\; and \;\; V = W/2 \;\; and \;\; k_{rect} = 1.2)$$

Integration limits are from 0 to L/2

$$= \left[(W/2)\,x\right]^2 / 2EI]\,dx \; + 2\int\left(3V^2/10GA\right) dx$$

$$= \int\left[W^2x^2/4EI\right] dx + \int\left(3W^2/10GA\right) dx$$

Integration limits are from 0 to L/2

2.  From Castigiano's method, the deflection of the point 'i' in the direction of the force $F_i$ (W in this case) will be found from the partial derivative.

$$yi = \partial U / \partial Fi = \partial U / \partial W$$

$$y_i = \int\left[2W\,x^2/4EI\right] dx \; + \; \int\left(6\,W/10GA\right) dx$$

$$y_i = \int\left(Wx^2/2EI\right) dx \; + \int\left(3W/5GA\right) dx$$

Integration limits are from 0 to L/2

$$y_i = WL^3/48EI \; + \; 3WL/10GA$$

$$= \left[40000 \times 2^3 / 48 \times 200 \times 1000 \times 417 \times 10^{-6}\right] + 3 \times 40000/10 \times 80 \times 1000 \times \left(0.1 \times 0.05\right)$$

$$= 7.8 + 0.060 \; = \; 7.860 \; mm$$

It is to be noted that deflection due to M is 7.8 mm while due to shear force is only 0.060 mm. Also finding of strain energy due to transverse shear is complex.

Therefore deflection due to transverse shear is quite small and hence normally not included in the calculations. So in a beam, find strain energy due to bending moment only and then find the deflection only due to the bending moment.

## Macaulay's Method

If the loading conditions change along the span of beam, there is corresponding change in moment equation. This requires that a separate moment equation be written between each change of load point and that two integration be made for each such moment equation. Evaluation of the constants introduced by each integration can become very involved. Fortunately, these complications can be avoided by writing single moment equation in such a way that it becomes continuous for entire length of the beam in spite of the discontinuity of loading.

For example consider the beam shown in fig below:

Let us write the general moment equation using the definition $M = \left(\sum M\right)_L$ Which means that we consider the effects of loads lying on the left of an exploratory section. The moment equations for the portions AB,BC and CD are written as follows:

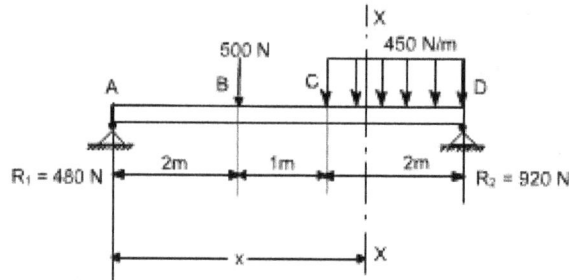

$$M_{AB} = 480 \times N.m$$

$$M_{BC} = \left[480 \times -500(x-2)\right] N.m$$

$$M_{CD} = \left[480 \times -500(x-2) - \frac{450}{2}(x-3)^2\right] N.m$$

It may be observed that the equation for $M_{CD}$ will also be valid for both MAB and MBC provided that the terms ( x - 2 ) and ( x - 3 )² are neglected for values of x less than 2 m and 3 m, respectively. In other words, the terms ( x - 2 ) and ( x - 3 )² are nonexistent for values of x for which the terms in parentheses are negative.

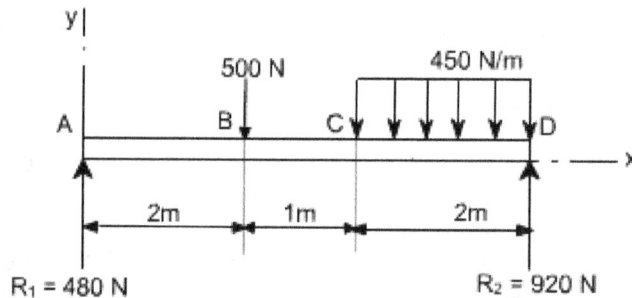

As an clear indication of these restrictions, one may use a nomenclature in which the usual form of parentheses is replaced by pointed brackets, namely, ‹ ’. With this change in nomenclature, we obtain a single moment equation.

$$M = \left(480 \times -500(x-2) - \frac{450}{2}(x-3)^2\right) N.m$$

Which is valid for the entire beam if we postulate that the terms between the pointed brackets do not exists for negative values; otherwise the term is to be treated like any ordinary expression.

As an another example, consider the beam as shown in the fig below. Here the distributed load extends only over the segment BC. We can create continuity, however, by assuming that the distributed load extends beyond C and adding an equal upward-distributed load to cancel its effect beyond C, as shown in the adjacent fig below. The general moment equation, written for the last segment DE in the new nomenclature may be written as:

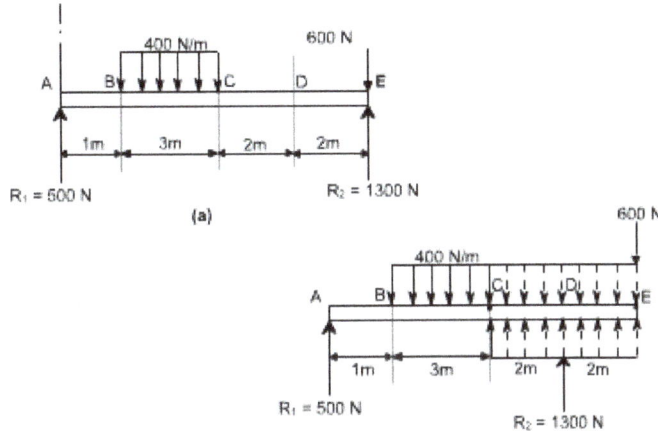
(a)

$$M = \left[500 \times -\frac{450}{2}(x-1)^2 + \frac{400}{2}(x-4)^2 + 1300(x-6)\right] N.m$$

It may be noted that in this equation effect of load 600 N won't appear since it is just at the last end of the beam so if we assume the exploratory just at section at just the point of application of 600 N than x = 0 or else we will here take the X - section beyond 600 N which is invalid.

## Procedure to Solve the Problems

(i) After writing down the moment equation which is valid for all values of 'x' i.e. containing pointed brackets, integrate the moment equation like an ordinary equation.

(ii) While applying the B.C's keep in mind the necessary changes to be made regarding the pointed brackets.

llustrative Examples :

1. A concentrated load of 300 N is applied to the simply supported beam as shown in Fig. Determine the equations of the elastic curve between each change of load point and the maximum deflection in the beam.

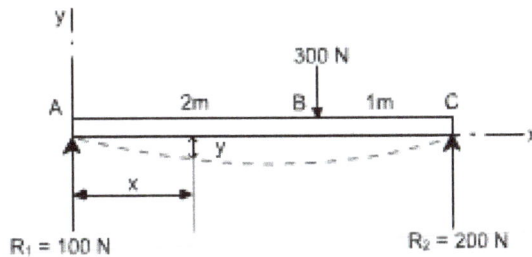

Solution : writing the general moment equation for the last portion BC of the loaded beam,

$$El\frac{d^2y}{dx^2} = M = \left(100x - 300\langle x - 2\rangle\right) N.m$$

Integrating twice the above equation to obtain slope and the deflection.

$$El\frac{dy}{dx} = \left(50x^2 - 150\langle x-2\rangle^2 + C_1\right)N.m^2$$

$$Ely = \left(\frac{50}{3}x^3 - 50\langle x-2\rangle^3 + C_1x + C_2\right)N.m^3$$

To evaluate the two constants of integration. Let us apply the following boundary conditions:

1. At point A where $x = 0$, the value of deflection $y = 0$. Substituting these values in Eq. $Ely =$ $\left(\frac{50}{3}x^3 - 50\langle x-2\rangle^3 + C_1x + C_2\right)N.m^3$ we find $C_2 = 0$.keep in mind that $< x -2 >^3$ is to be neglected for negative values.

2. At the other support where $x = 3m$, the value of deflection $y$ is also zero.

substituting these values in the deflection Eq. $Ely = \left(\frac{50}{3}x^3 - 50\langle x-2\rangle^3 + C_1x + C_2\right)N.m^3$, we obtain,

$$0 = \left(\frac{50}{3}3^3 - 50(3-2)^3 + 3.C_1\right) \text{ or } C_1 = -133\,N.m^2$$

Having determined the constants of integration, let us make use of Eqs.

$$El\frac{dy}{dx} = \left(50x^2 - 150\langle x-2\rangle^2 + C_1\right)N.m^2 \text{ and } Ely = \left(\frac{50}{3}x^3 - 50\langle x-2\rangle^3 + C_1x + C_2\right)N.m^3 \text{ to}$$

rewrite the slope and deflection equations in the conventional form for the two portions.

Sagment $AB(0 \le x \le 2m)$

$$El\frac{dy}{dx} = \left(50x^2 - 133\right)N.m^2$$

$$Ely = \left(\frac{50}{3}x^3 - 133x\right)N.m^3$$

Sagment $BC(2m \le x \le 3m)$

$$El\frac{dy}{dx} = \left(50x^2 - 150(x-2)^2 - 133x\right)N.m^2$$

$$Ely = \left(\frac{50}{3}x^3 - 50(x-2)^3 - 133x\right)N.m^3$$

Continuing the solution, we assume that the maximum deflection will occur in the segment AB. Its location may be found by differentiating Eq. $Ely = \left(\frac{50}{3}x^3 - 133x\right)N.m^3$ with respect to x and setting the derivative to be equal to zero, or, what amounts to the same thing, setting the slope equation $EI\dfrac{dy}{dx}=\left(50x^2-133\right)N.m^2$ equal to zero and solving for the point of zero slope.

We obtain,

$50x^2-133=0$ or $x=1.63$ m (It may be kept in mind that if the solution of the equation does not yield a value < 2 m then we have to try the other equations which are valid for segment BC).

Since this value of x is valid for segment AB, our assumption that the maximum deflection occurs in this region is correct. Hence, to determine the maximum deflection, we substitute $x=1.63$ m in

Eq $EIy=\left(\dfrac{50}{3}x^3-133x\right)N.m^3$, which yields,

$$EIy\big|_{max}\,m=-145\,N.m^3$$

The negative value obtained indicates that the deflection y is downward from the x axis. quite usually only the magnitude of the deflection, without regard to sign, is desired; this is denoted by d, the use of y may be reserved to indicate a directed value of deflection.

if $E=30\,Gpa$ and $I=1.9\times10^6\,mm^4=1.9\times10^{-6}\,m^4$, Eq. (h) becomes

Then $y\big|_{max}\,m=\left(30\times10^9\right)\left(1.9\times10^{-6}\right)$

$$=-2.54\,mm$$

Example:

It is required to determine the value of EIy at the position midway between the supports and at the overhanging end for the beam shown in figure below.

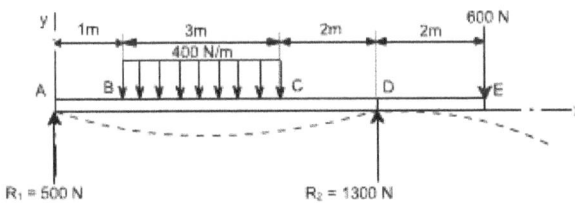

Solution:

Writing down the moment equation which is valid for the entire span of the beam and applying the differential equation of the elastic curve, and integrating it twice, we obtain,

$$EI\dfrac{d^2y}{dx^2}=M=\left(500x-\dfrac{400}{2}(x-1)^2+\dfrac{400}{2}(x-4)^2+1300(x-6)\right)N.m$$

$$EI\dfrac{dy}{dx}=\left(250x^2-\dfrac{200}{3}(x-1)^3+\dfrac{200}{3}(x-4)^3+650(x-6)^2+C_1\right)N.m$$

$$Ely = \left(\frac{250}{3}x^3 - \frac{50}{3}(x-1)^4 + \frac{50}{3}(x-4)^4 + \frac{650}{3}(x-6)^3 + C_1 x + C_1\right) N.m^3$$

To determine the value of $C_2$, It may be noted that $Ely = 0$ at $x = 0$, which gives $C_2 = 0$. Note that the negative terms in the pointed brackets are to be ignored Next, let us use the condition that $Ely = 0$ at the right support where $x = 6m$. This gives,

$$0 = \frac{250}{3}(6)^3 - \frac{50}{3}(5)^4 + \frac{50}{3}(2)^4 + 6C_1 \quad \text{or} \quad C_1 = -1308 N.m^2$$

Finally, to obtain the midspan deflection, let us substitute the value of $x = 3m$ in the deflection equation for the segment BC obtained by ignoring negative values of the bracketed terms $\langle x-4\rangle^4$ and $\langle x-6\rangle^3$. We obtain,

$$Ely = \frac{250}{3}(3)^3 - \frac{50}{3}(2)^4 - 1308(3) = -1941 N.m^3$$

For the overhanging end where x=8m, we have,

$$Ely = \left(\frac{250}{3}(8)^3 - \frac{50}{3}(7)^4 + \frac{50}{3}(4)^4 + \frac{650}{3}(2)^3 - 1308(8)\right)$$

$$= -1814 N.m^3$$

Example:

A simply supported beam carries the triangularly distributed load as shown in figure. Determine the deflection equation and the value of the maximum deflection.

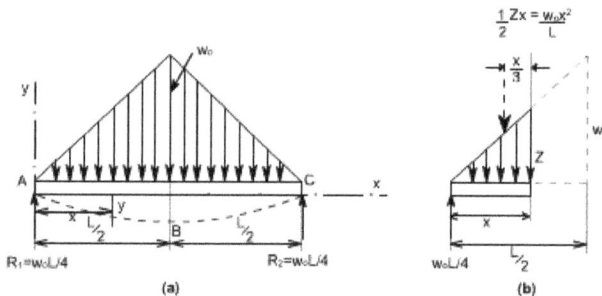

(a)　　　　　　　　　　(b)

Solution:

Due to symmetry, the reactions is one half the total load of $1/2 w_0 L$, or $R_1 = R_2 = 1/4 w_0 L$. Due to the advantage of symmetry to the deflection curve from A to B is the mirror image of that from C to B. The condition of zero deflection at A and of zero slope at B do not require the use of a general moment equation. Only the moment equation for segment AB is needed and this may be easily written with the aid of figure (b).

Taking into account the differential equation of the elastic curve for the segment AB and integrating twice, one can obtain,

$$E1\frac{d^2y}{dx^2} = M_{AB} = \frac{w_oL}{4}x - \frac{w_ox^2}{L}\cdot\frac{x}{3}$$

$$EI\frac{dy}{dx} = \frac{w_oLx^2}{8} - \frac{w_ox^4}{12L} + C_1$$

$$EIy = \frac{w_oLx^3}{24} - \frac{w_ox^5}{60L} + C_1x + C_2$$

In order to evaluate the constants of integration, let us apply the B.C's we note that at the support

A, $y = 0$ at $x = 0$. Hence from equation $EIy = \dfrac{w_oLx^3}{24} - \dfrac{w_ox^5}{60L} + C_1x + C_2$, we get $C2 = 0$. Also, be-

cause of symmetry, the slope $dy/dx = 0$ at midspan where $x = L/2$. Substituting these conditions in

equation $EI\dfrac{dy}{dx} = \dfrac{w_oLx^2}{8} - \dfrac{w_ox^4}{12L} + C_1$ we get,

$$0 = \frac{w_oL}{8}\left(\frac{L}{2}\right)^2 - \frac{w_o}{12L}\left(\frac{L}{2}\right) + C_1C_2 = -\frac{5w_oL^3}{192}$$

Hence the deflection equation from A to B (and also from C to B because of symmetry) becomes,

$$EIy = \frac{w_oLx^3}{24} - \frac{w_ox^5}{60L} - \frac{5w_oL^3x}{192}$$

Which reduces to

$$EIy = -\frac{w_ox}{960L}\left(25L^4 - 40L^2x^2 + 16x^4\right)$$

The maximum deflection at midspan, where $x=L/2$ is then found to be,

$$EIy = -\frac{w_oL^4}{120}$$

Example: couple acting

Consider a simply supported beam which is subjected to a couple M at a distance 'a' from the left end. It is required to determine using the Macauley's method.

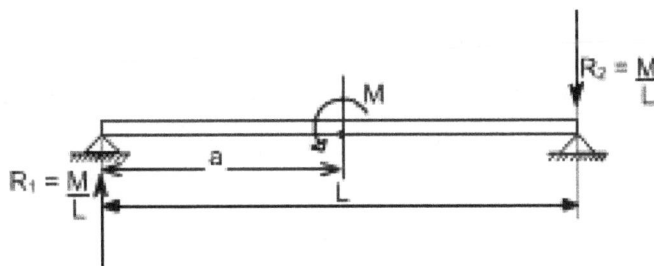

To deal with couples, only thing to remember is that within the pointed brackets we have to take some quantity and this should be raised to the power zero.i.e. $M\langle x-a\rangle^{0}$. We have taken the power o (zero) ' because ultimately the term $M\langle x-a\rangle^{0}$ Should have the moment units.Thus with integration the quantity $\langle x-a\rangle$ becomes either $\langle x-a\rangle^{1}$ or $\langle x-a\rangle^{2}$

Or

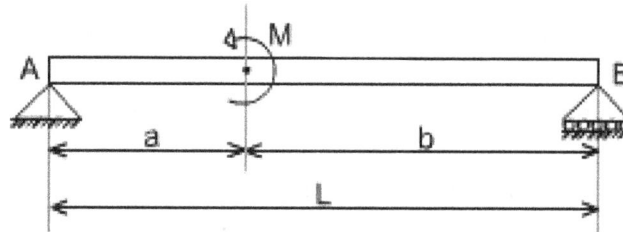

Therefore, writing the general moment equation we get,

$$M = R_1 x - M\langle x-a\rangle \text{ or } El\frac{d^2y}{dx^2} = M$$

Integrating twice we get,

$$El\frac{dy}{dx} = R_1 . \frac{x^2}{2} - M\langle x-a\rangle^{1} + C_1$$

$$Ely = R_1 . \frac{x^3}{6} - \frac{M}{2}\langle x-a\rangle^{2} + C_1 x + C_2$$

Example:

A simply supported beam is subjected to U.d.l in combination with couple M. It is required to determine the deflection.

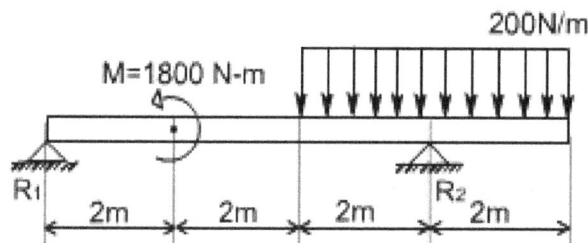

This problem may be attempted in the same way. The general moment equation my be written as,

$$M(x) = R_1 x - 1800\langle x-2\rangle^{0} - \frac{200\langle x-4\rangle\langle x-4\rangle}{2} + R_2\langle x-6\rangle$$

$$= R_1 x - 1800\langle x-2\rangle^{0} - \frac{200\langle x-4\rangle^{2}}{2} + R_2\langle x-6\rangle$$

Thus

$$\text{El}\frac{d^2y}{dx^2} = R_1 x - 1800\langle x\text{-}2\rangle^0 - \frac{200\langle x\text{-}4\rangle^2}{2} + R_2\langle x\text{-}6\rangle$$

Integrate twice to get the deflection of the loaded beam.

## Direct Stiffness Method

Stiffness method is useful in analysing the structures like beams and frames, which are indeterminate, having kinematic indeterminacy lesser than the static indeterminacy. Number of unknown displacements should be more than the unknown reactions to apply stiffness based methods. When the building to be studied becomes complex with number of members exceeding three, ordinary stiffness based methods become cumbersome for performing hand calculations.

### Direct Stiffness Method

Direct stiffness method is method founded on stiffness matrices written based on member configurations, and how the individual members and nodes deform (rotate/deflect) behave upon loading.

Local stiffness matrix is formed for individual members, and assembled to form a global stiffness matrix K for the whole structure.

Stiffness matrix k is written differently in case of axial members (trusses), bending members (beams and frames).

Example:

Consider the example of 3-member truss for demonstrating the procedure followed in case of direct stiffness method.

Consider the case of 2-D truss members, in which the joints can undergo horizontal and vertical displacements (2 degrees of freedom). A typical truss member has 2 nodes with 4 degrees of freedom.

An axial member will have local stiffness matrix of size 4×4.

$$[k] = \frac{EA}{L}\begin{bmatrix} \cos^2\theta & \cos\theta\sin\theta & -\cos^2\theta & -\cos\theta\sin\theta \\ \cos\theta\sin\theta & \sin^2\theta & -\cos\theta\sin\theta & -\sin^2\theta \\ -\cos^2\theta & -\cos\theta\sin\theta & \cos^2\theta & \cos\theta\sin\theta \\ -\cos\theta\sin\theta & -\sin^2\theta & \cos\theta\sin\theta & \sin^2\theta \end{bmatrix}$$

Here, $\theta$ is the inclination of truss member with horizontal axis A is the Area of member, L is length of member, and E is the elastic modulus of member.

Global stiffness matrix is to be gathered from local matrix (stiffness) of all elements, by adding up element corresponding to suitable rows and columns. If the truss has 3 elements, the global matrix of stiffness will be of size 6×6.

Load displacement connection matrix is as below.

$$[p] = [k][u]$$

Here, [p] is the load Matrix of size 6×1 and [u] is the joint displacement matrix of 6×1 size in respective directions.

Joint displacements depend on support conditions, if hinged displacements in both direction are zero, and if the joint is normal truss joint, displacements are unknown.

Using the zero displacement nodes, compute the reactions.

Then proceed to compute the displacement of other nodes with known point loads acting on other joints.

Force in each member should be computed using the relation below.

$$[p'] = [k'][u']$$

$$[k'] = \frac{AE}{L}[\cos\theta \quad \sin\theta \quad -\cos\theta \quad -\sin\theta]$$

Here, [p'] is joint load matrix for single member of size 2×1, [k'] is matrix of size 1×4, and [u'] is the calculated joint displacement matrix of size 2×1.

Thus, it is used for trusses with size more than 3, solved for forces writing computer programs with the formulation.

## Limit State Design

Limit state design (LSD) refers to a structural engineering design method.

A degree of loading or other actions imposed on a structure can result in a 'limit state', where the structure's condition no longer fulfils its design criteria, such as; fitness for use, structural integrity, durability, and so on. Limit states are conditions of potential failure.

All actions likely to occur during a structure's design life are considered during the LSD method, to ensure that the structure remains fit for use with appropriate levels of reliability.

LSD involves estimating the subjected loads on a structure, choosing the sizes of members to check, and selecting the appropriate design criteria. LSD requires two principal criteria to be satisfied: the ultimate limit state (ULS) and the serviceability limit state (SLS).

## Ultimate Limit State (ULS)

The ultimate limit state is the design for the safety of a structure and its users by limiting the stress that materials experience. In order to comply with engineering demands for strength and stability under design loads, ULS must be fulfilled as an established condition.

The ULS is a purely elastic condition, usually located at the upper part of its elastic zone (approximately 15% lower than the elastic limit). This is in contrast to the ultimate state (US) which involves excessive deformations approaching structural collapse, and is located deeply within the plastic zone.

If all factored bending, shear and tensile or compressive stresses are below the calculated resistances then a structure will satisfy the ULS criterion. Safety and reliability can be assumed as long as this criterion is fulfilled, since the structure will behave in the same way under repetitive loadings.

## Serviceability Limit State (SLS)

The serviceability limit state is the design to ensure a structure is comfortable and useable. This includes vibrations and deflections (movements), as well as cracking and durability. These are the conditions that are not strength-based but still may render the structure unsuitable for its intended use, for example, it may cause occupant discomfort under routine conditions.

It might also involve limits to non-structural issues such as acoustics and heat transmission.

SLS requirements tend to be less rigid than strength-based limit states as the safety of the structure is not in question. A structure must remain functional for its intended use subject to routine loading in order to satisfy SLS criterion.

Structures are designed according to a standardized code that have properties (or specific criteria) that need to be met during the design procedure. The limit state of a structure is when the structure goes beyond the specified criteria and it 'breaks'. Limit state design can therefore be defined as the process of designing a structure so that it doesn't break and remains fit for its designed use.

There are numerous design philosophies used by civil engineers. The three most prominent, as found through research, will be briefly discussed. These are:

1.  Working Stress Method (WSM)

2.  Ultimate Load Method (ULM)

3.  Limit State Method (LSM)

The Working Stress Method assumes that all material used in the design behaves in a linear elastic manner and calculations are based on service conditions. The expected working loads on the structure are induced as stresses on the structure during the design and these are restricted (to be below the permissible stresses) to ensure adequate safety. The factor of safety is the ratio of the strength of the material to the permissible stress. However, as working loads cannot be kept within the permissible stresses this method is not always viable.

The assumption in the Ultimate Load Method is that the material is not linear-elastic meaning that different loads can have different safety factors. Calculations are based on ultimate load conditions and the stress conditions at the site of failure or impending collapse is analysed. This results in more slender (therefore more economic!) sections. However, this method can result in cracks and deflections that are excessive due to the serviceability properties not being met.

In the Limit State Method (aka as Limit State Design), the design of the structure is considered for both the serviceability and ultimate load state. This is therefore the better design philosophy to employ. The two different limit states are the Ultimate Limit State (ULS) and the Serviceability Limit State (SLS), for which distinction are made in the code SANS10160: 5.1.3... These states each have their own minimum level of reliability - which is colloquially known as the Safety Index.

# Moment Distribution Method

Moment distribution is a great method for quickly computing end moments on continuous beams. Over the years, several variations of the method have been presented. These methods take advantage of various observations made about the process. While this method can be applied to a variety of indeterminate structures, the discussion here is limited to continuous beams that may or may not be continuous with supporting columns.

The moment distribution method begins by determining the relative flexural stiffness, in the plane of loading, of all the elements rigidly connected to each joint. Rotational stiffness of a member is proportional to material and geometric stiffness:

Relative Rotational Stiffness = EI/L

Where:

- E is the modulus of elasticity of the material,

- I is the moment of inertia about the axis of bending in the plane of the frame, and

- L is the length of the member.

The basic premise of the moment distribution method is that any unbalanced moment on a joint is redistributed to the members rigidly attached to that joint in proportion to contributions each element makes to the total rotational stiffness of the joint. Consequently, distribution factors are computed at each joint, one for each element attached to the joint that provides rotational stiffness in the considered plane. The equation for computing the distribution factor of the $i^{th}$ member at a joint with n members is:

$$DF_i = \frac{\left(EI/L\right)_i}{\sum_{j=1}^{j=n}\left(EI/L\right)_j}$$

Note that if any of the variables is constant for all members of the joint, then the constant variable will cancel out of the equation.

The distribution factors are computed joint by joint for the ends of each member connected to the joint.

The moment distribution method assumes that all joints are "fixed" at the start of the problem. This means that when loads are applied to each span, fixed end moments are developed at each end of the loaded span. The fixed end moments are a function of the nature and location of the applied loads on the span. If there is more than one load source on a span, super position can be used to determine the total fixed end moment at the end of each span.

Concerning notation, in this method the end moments are considered to be reactions. As a result, the right hand rule is used to determine the sign of the moments. Do not use beam internal force notation.

After computing the fixed end moments (FEM) for each span, you will observe that there is an unbalanced moment at each joint. In other words, the sum of moments at the joint does not equal zero, a necessary requirement for equilibrium. To rectify this problem, a joint is "released" and the unbalanced moment (i.e. the difference of the FEMs) is divided among the members attached to joint in proportion to their contributions to the joint rotational stiffness using the distribution factors. Once the distributed moments are added to each pre-existing FEM at the joint, the resulting FEMs are "balanced". In other words, the sum of the moments at the joint equal zero and equilibrium is satisfied at that joint (for now!).

The addition of moment to each element attached to the joint induces a moment on the opposite end of each member. This is called "carry over" (CO). The carry over moment is added to the pre-existing FEM at that joint. The CO moment tends to unbalance the joints adjacent to newly balanced joint. These joints are, in turn released, balanced and send back CO moments to unbalance the adjacent joint.

Joints are successively released and balanced until the CO moments get small enough to ignore. This means that this is an iterative process that looks for convergence.

An example may help to illustrate the process.

Consider the beam shown in figure. This beam will have a constant E and I for all three spans, so the relative stiffness of each can be computed as 1/L.

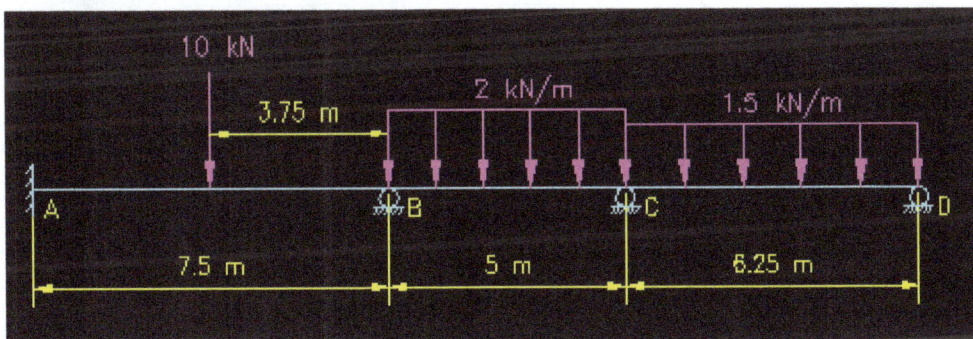
Figure: Beam Problem Definition

## Compute the Distribution Factors

For Joint "A":

Two items contribute to the rotational stiffness at A. One is the beam AB the other is the infinitely stiff support. The distribution factor at the left end of beam AB then is:

$$DF_{AB,L} = (1/7.5) / (\text{infinity} + 1/7.5) = 0$$

The distribution factor for the support at A is:

$$DF_A = \text{infinity} / (\text{infinity} + 1/7.5) = 1$$

For Joint "B":

The two items contributing to rotation stiffness at B are the beams that come in from either side. The support is a roller so it does not contribute to the rotational support of the joint. The two distribution factors are:

$$DF_{AB,R} = (1/7.5) / ( (1/7.5) + (1/5) ) = 0.400$$

$$DF_{BC,L} = (1/5) / ( (1/7.5) + (1/5) ) = 0.600$$

For Joint "C":

The two items contributing to rotation stiffness at C are the beams that come in from either side. The support is a roller so it does not contribute to the rotational support of the joint. The two distribution factors are:

$$DF_{BC,R} = (1/5) / ( (1/6.25) + (1/5) ) = 0.556$$

$$DF_{CD,L} = (1/6.25) / ( (1/6.25) + (1/5) ) = 0.444$$

For Joint "D":

There is only one item contributing to rotation stiffness at D: Beam CD. The support is a roller so it does not contribute to the rotational support of the joint. The two distribution factors are:

$$DF_{CD,R} = (1/6.25) / ( (1/6.25) + (0) ) = 1.000$$

$$DF_D = (0) / ( (1/6.25) + (0) ) = 0.000$$

Table summarizes the progress to this point.

## Distribution Factors

|       | Span AB  |       | Span BC |       | Span CD |       | Units |
|-------|----------|-------|---------|-------|---------|-------|-------|
| Span  | 7.5      |       | 5       |       | 6.25    |       | m     |
| 1/L   | 0.133333 |       | 0.2     |       | 0.16    |       |       |
| Joint | A        | B     | B       | C     | C       | D     |       |
| DF    | 0.000    | 0.400 | 0.600   | 0.556 | 0.444   | 1.000 |       |

## Compute the Fixed End Moments

The fixed end moments are computed span by span using the given loading for each span and considering the ends of the beam to be fixed.

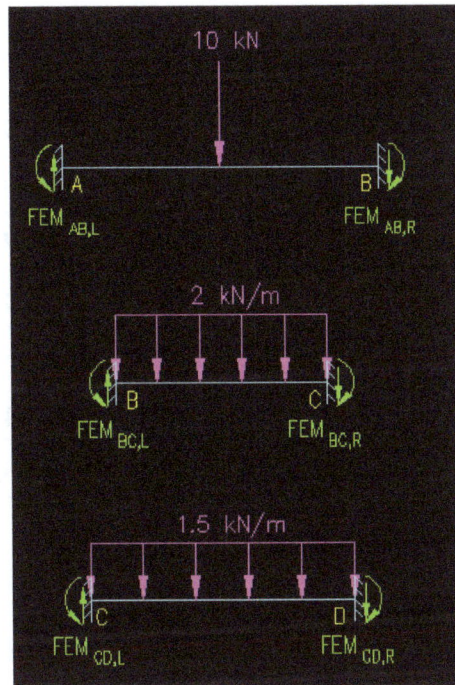

Fixed End Moments

Any number of sources can provide equations for fixed end moments. Maybe, we'll eventually add a table of them here. For now, you can use the beam tables found in the Steel Construction Manual, the Fundamentals of Engineering (FE) exam reference manual, or derive your own.

Figure shows each of the spans, their loading, and the definition of the FEMs. Note that each of the right end FEMs are right hand rule negative moments. It is important to keep the sign.

For Span AB, the FEMS are:

$$FEM_{AB,L} = + PL / 8 = + (10 \text{ kN}) (7.5 \text{ m}) / 8 = + 9.375 \text{ kN/m}$$

$$FEM_{AB,R} = - PL / 8 = - (10 \text{ kN}) (7.5 \text{ m}) / 8 = - 9.375 \text{ kN/m}$$

The two FEMs have the same magnitude since the load is centered on the span. They would be different otherwise.

For Span BC, the FEMS are:

$$FEM_{BC,L} = + wL^2 / 12 = + (2 \text{ kN/m}) (5 \text{ m})^2 / 12 = + 4.167 \text{ kN/m}$$

$$FEM_{BC,L} = + wL^2 / 12 = - (2 \text{ kN/m}) (5 \text{ m})^2 / 12 = - 4.167 \text{ kN/m}$$

For Span CD, the FEMS are:

$$FEM_{CD,L} = + wL^2 / 12 = + (1.5 \text{ kN/m}) (6.25 \text{ m})^2 / 12 = + 4.883 \text{ kN/m}$$

$$FEM_{CD,L} = + wL^2 / 12 = - (1.5 \text{ kN/m}) (6.25 \text{ m})^2 / 12 = - 4.883 \text{ kN/m}$$

Table show our spread sheet table expanded to include the initial FEMs.

## Distribution Factors

|  | Span AB |  | Span BC |  | Span CD |  | Units |
|---|---|---|---|---|---|---|---|
| Span | 7.5 |  | 5 |  | 6.25 |  | m |
| 1/L | 0.133333 |  | 0.2 |  | 0.16 |  |  |
| Load | 10 | kN | 2 | kN/m | 1.5 | kN/m |  |
| Joint | A | B | B | C | C | D |  |
| DF | 0.000 | 0.400 | 0.600 | 0.556 | 0.444 | 1.000 |  |
| FEM | 9.375 | -9.375 | 4.167 | -4.167 | 4.883 | -4.883 | kN-m |

## Balancing the Joints and Doing the Carry Over

At this time, all joints are said to be fixed. In other words, it is as if an invisible lock has taken the imbalance at each joint. This explains why equilibrium is not satisfied at a joint. Let's take each joint and do a balance.

Joint A:

> The table that we are developing does not show the support side of the joint. This is common. Joint A is always balanced because the fixed support moment matches (with opposite sign) the beam end moment. This joint does not require balancing.

Joint B:

> The unbalanced moment a this joint equals (-9.375 kN-m + 4.167 kN-m) = - 5.208 kN-m. To balance this joint we need to add + 5.208 kN-m to the joint. This is done by adding to each beam end in proportion to its contribution to the rotational stiffness of the joint. This is what the distribution factors are for.

The new FEMs are:

> $FEM_{AB,R}$ = - 9.375 kN/m + $DF_{AB,R}$ (+5.208 kN-m)
>
> $FEM_{AB,R}$ = - 9.375 kN/m + 0.400 (+5.208 kN-m) = -9.375 kN-m + *2.083 kN-m*
>
> $FEM_{AB,R}$ = - 7.292 kN/m
>
> $FEM_{BC,L}$ = + 4.167 kN/m + $DF_{BC,L}$ (+5.208 kN-m)
>
> $FEM_{BC,L}$ = + 4.167 kN/m + (0.600) (+5.208 kN-m) = + 4.167 kN/m + *3.125 kN-m*
>
> $FEM_{BC,L}$ = + 7.292 kN/m

Note that the sum of the two FEMs is zero. The joint is balanced.

The result of the balancing creates moments at the far ends of Beams AB and BC equal to half the balancing moments. Table shows the balancing moments at joint B and the resulting carry over moments at joints A and C. The summation line shows the sum of the FEM, balance, and CO lines to show the current end moments on each span. Note that only joint B is "balanced" at this point.

## Joint B Balance and Carry Over

|  | Span AB |  | Span BC |  | Span CD |  | Units |
|---|---|---|---|---|---|---|---|
| Span | 7.5 |  | 5 |  | 6.25 |  | m |
| 1/L | 0.133333 |  | 0.2 |  | 0.16 |  |  |
| Load | 10 | kN | 2 | kN/m | 1.5 | kN/m |  |
| Joint | A | B | B | C | C | D |  |
| DF | 0.000 | 0.400 | 0.600 | 0.556 | 0.444 | 1.000 |  |
| FEM | 9.375 | -9.375 | 4.167 | -4.167 | 4.883 | -4.883 | kN-m |
| Balance | 0.000 | 2.083 | 3.125 |  |  |  | kN-m |
| CO | 1.042 | 0.000 |  | 1.563 |  |  | kN-m |
| Sum | 10.417 | -7.292 | 7.292 | -2.604 | 4.883 | -4.883 | kN-m |

Joint C:

Ignoring the carry over moment from the prior joint, the unbalanced moment at this joint equals (-4.167 kN-m + 4.883 kN-m) = + 0.716 kN-m. To balance this joint we need to add - 0.716 kN-m to the joint. This is done by adding to each beam end in proportion to its contribution to the rotational stiffness of the joint. This is what the distribution factors are for.

The new FEMs are:

$$FEM_{BC,R} = -4.167 \text{ kN/m} + DF_{BC,R} (-0.716 \text{ kN-m})$$

$$FEM_{BC,R} = -4.167 \text{ kN/m} + 0.556 (-0.716 \text{ kN-m}) = -4.167 \text{ kN-m} - 0.398 \text{ kN-m}$$

$$FEM_{BC,R} = -4.565 \text{ kN/m}$$

$$FEM_{CD,L} = +4.883 \text{ kN/m} + DF_{CD,L} (-0.716 \text{ kN-m})$$

$$FEM_{CD,L} = +4.883 \text{ kN/m} + (0.444) (-0.716 \text{ kN-m}) = +4.883 \text{ kN/m} - 0.318 \text{ kN-m}$$

$$FEM_{BC,L} = +4.565 \text{ kN/m}$$

Note that the sum of the two FEMs is zero. The joint is balanced, if you continue to ignore the carryover from joint B. Table shows the balancing moments and carry over moments (highlighted in yellow) resulting from this operation.

## Joint B Balance and Carry Over

|  | Span AB |  | Span BC |  | Span CD |  | Units |
|---|---|---|---|---|---|---|---|
| Span | 7.5 |  | 5 |  | 6.25 |  | m |
| 1/L | 0.133333 |  | 0.2 |  | 0.16 |  |  |
| Load | 10 | kN | 2 | kN/m | 1.5 | kN/m |  |

| Joint | A | B | B | C | C | D | |
|-------|-----|-----|-----|-----|-----|-----|-----|
| DF | 0.000 | 0.400 | 0.600 | 0.556 | 0.444 | 1.000 | |
| FEM | 9.375 | -9.375 | 4.167 | -4.167 | 4.883 | -4.883 | kN-m |
| Balance | 0.000 | 2.083 | 3.125 | -0.398 | -0.318 | | kN-m |
| CO | 1.042 | 0.000 | -0.199 | 1.563 | | -0.159 | kN-m |
| Sum | 10.417 | -7.292 | 7.093 | -3.002 | 4.565 | -5.042 | kN-m |

Joint D:

As we did for Joint C, we ignore the carry over moment from the prior joint. As we did at joint A, we did not include support moment in the table. For this case the joint is a pin, so no moment can be there when we end. The unbalanced moment a this joint equals -4.883 kN-m. To balance this joint we need to add +4.883 kN-m to the joint. This is done by adding to each beam end in proportion to its contribution to the rotational stiffness of the joint. This is what the distribution factors are for.

The new FEM is:

$$FEM_{CD,R} = -4.883 \text{ kN/m} + DF_{CD,R} (+4.883 \text{ kN-m})$$

$$FEM_{CD,R} = -4.883 \text{ kN/m} + 1.000 (+4.883 \text{ kN-m}) = -4.167 \text{ kN-m} + 4.883 \text{ kN-m}$$

$$FEM_{CD,R} = 0 \text{ kN/m}$$

Note that the FEM is zero. The joint is balanced, if you continue to ignore the carry-over from joint C. Table shows the balancing moment and carry over moment (highlighted in yellow) resulting from this operation.

## Joint B Balance and Carry Over

| | Span AB | | Span BC | | Span CD | | Units |
|-------|---------|-----|---------|-----|---------|-----|-------|
| Span | 7.5 | | 5 | | 6.25 | | m |
| 1/L | 0.133333 | | 0.2 | | 0.16 | | |
| Load | 10 | kN | 2 | kN/m | 1.5 | kN/m | |
| Joint | A | B | B | C | C | D | |
| DF | 0.000 | 0.400 | 0.600 | 0.556 | 0.444 | 1.000 | |
| FEM | 9.375 | -9.375 | 4.167 | -4.167 | 4.883 | -4.883 | kN-m |
| Balance | 0.000 | 2.083 | 3.125 | -0.398 | -0.318 | 4.883 | kN-m |
| CO | 1.042 | 0.000 | -0.199 | 1.563 | 2.441 | -0.159 | kN-m |
| Sum | 10.417 | -7.292 | 7.093 | -3.002 | 7.006 | -0.159 | kN-m |

Adding the original FEMs to the corresponding balancing moments and carry over moments we arrive at a new FEM state. Notice that the imbalance at each joint is significantly less than what the original FEMs provided. We are starting to converge on the solution.

At this point we repeat the balance, carry over and sum steps until the imbalances become small enough for our purposes. Generally, this occurs when the unbalanced moments are about 1% of the end moments. With a spread sheet, repeating the steps is simply a matter of copying the balance, carry over and sum steps once the formulas are all set up. Table shows the results of nine balancing cycles.

## The Full Solution

|         | Span AB  |       | Span BC |        | Span CD |        | Units |
|---------|----------|-------|---------|--------|---------|--------|-------|
| Span    | 7.5      |       | 5       |        | 6.25    |        | m     |
| 1/L     | 0.133333 |       | 0.2     |        | 0.16    |        |       |
| Load    | 10       | kN    | 2       | kN/m   | 1.5     | kN/m   |       |
| Joint   | A        | B     | B       | C      | C       | D      |       |
| DF      | 0.000    | 0.400 | 0.600   | 0.556  | 0.444   | 1.000  |       |
| FEM     | 9.375    | -9.375| 4.167   | -4.167 | 4.883   | -4.883 | kN-m  |
| Balance | 0.000    | 2.083 | 3.125   | -0.398 | -0.318  | 4.883  | kN-m  |
| CO      | 1.042    | 0.000 | -0.199  | 1.563  | 2.441   | -0.159 | kN-m  |
| Sum     | 10.417   | -7.292| 7.093   | -3.002 | 7.006   | -0.159 | kN-m  |
| Balance | 0.000    | 0.080 | 0.119   | -2.224 | -1.780  | 0.159  | kN-m  |
| CO      | 0.040    | 0.000 | -1.112  | 0.060  | 0.080   | -0.890 | kN-m  |
| Sum     | 10.456   | -7.212| 6.100   | -5.167 | 5.306   | -0.890 | kN-m  |
| Balance | 0.000    | 0.445 | 0.667   | -0.077 | -0.062  | 0.890  | kN-m  |
| CO      | 0.222    | 0.000 | -0.039  | 0.334  | 0.445   | -0.031 | kN-m  |
| Sum     | 10.679   | -6.767| 6.729   | -4.910 | 5.689   | -0.031 | kN-m  |
| Balance | 0.000    | 0.015 | 0.023   | -0.433 | -0.346  | 0.031  | kN-m  |
| CO      | 0.008    | 0.000 | -0.216  | 0.012  | 0.015   | -0.173 | kN-m  |
| Sum     | 10.687   | -6.752| 6.535   | -5.331 | 5.358   | -0.173 | kN-m  |
| Balance | 0.000    | 0.087 | 0.130   | -0.015 | -0.012  | 0.173  | kN-m  |
| CO      | 0.043    | 0.000 | -0.008  | 0.065  | 0.087   | -0.006 | kN-m  |
| Sum     | 10.730   | -6.665| 6.658   | -5.282 | 5.433   | -0.006 | kN-m  |
| Balance | 0.000    | 0.003 | 0.005   | -0.084 | -0.067  | 0.006  | kN-m  |
| CO      | 0.002    | 0.000 | -0.042  | 0.002  | 0.003   | -0.034 | kN-m  |
| Sum     | 10.731   | -6.662| 6.620   | -5.363 | 5.369   | -0.034 | kN-m  |
| Balance | 0.000    | 0.017 | 0.025   | -0.003 | -0.002  | 0.034  | kN-m  |
| CO      | 0.008    | 0.000 | -0.001  | 0.013  | 0.017   | -0.001 | kN-m  |
| Sum     | 10.740   | -6.645| 6.644   | -5.354 | 5.383   | -0.001 | kN-m  |
| Balance | 0.000    | 0.001 | 0.001   | -0.016 | -0.013  | 0.001  | kN-m  |
| CO      | 0.000    | 0.000 | -0.008  | 0.000  | 0.001   | -0.007 | kN-m  |
| Sum     | 10.740   | -6.645| 6.637   | -5.370 | 5.371   | -0.007 | kN-m  |
| Balance | 0.000    | 0.003 | 0.005   | -0.001 | 0.000   | 0.007  | kN-m  |
| CO      | 0.002    | 0.000 | 0.000   | 0.002  | 0.003   | 0.000  | kN-m  |
| Sum     | 10.742   | -6.642| 6.641   | -5.368 | 5.373   | 0.000  | kN-m  |

Notice that the changes to the end moments have become "small". It is time to stop.

With these end moments, the remaining shears and moments can be found using equilibrium equations.

# Slope Deflection Method

The slope-deflection method can be used to analyse statically determinate and indeterminate beams and frames. In this method it is assumed that all deformations are due to bending only. In other words deformations due to axial forces are neglected.

Consider a typical span of a continuous beam AB as shown in figure below. The beam has constant flexural rigidity EIand is subjected to uniformly distributed loading and concentrated loads as shown in the figure. The beam is kinematically indeterminate to second degree. In this lesson, the slope-deflection equations are derived for the simplest case i.e. for the case of continuous beams with unyielding supports. In the next lesson, the support settlements are included in the slope-deflection equations.

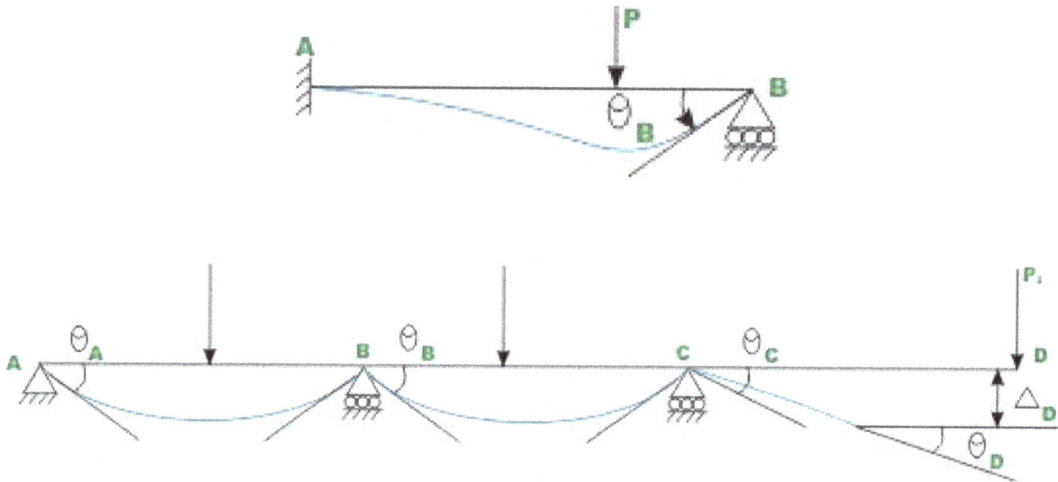

For this problem, it is required to derive relation between the joint end moments $M_{AB}$ and $M_{BA}$ in terms of joint rotations and loads acting on the beam. Two subscripts are used to denote end moments. For example, end moments $M_{AB}$ denote moment acting at joint A of the member AB. Rotations of the tangent to the elastic curve are denoted by one subscript. Thus, $\theta_A$ denotes the rotation of the tangent to the elastic curve at A. The following sign conventions are used in the slope-deflection equations:

- Moments acting at the ends of the member in counter clockwise direction are taken to be positive.

- The rotation of the tangent to the elastic curve is taken to be positive when the tangent to the elastic curve has rotated in the counter clockwise direction from its original direction.

The slope-deflection equations are derived by superimposing the end moments developed due to applied loads, rotation $\theta_A$ and rotation $\theta_B$. This condition is obtained by modifying the support conditions to fixed so that the unknown joint rotations become zero. The structure shown in figure is known as kinematically determinate structure or restrained structure. For this case, the end moments are denoted by $M_{AB}^F$ and $M_{BA}^F$. The fixed end moments are evaluated by force–method of analysis as discussed in the previous module. For example for fixed- fixed beam subjected to uniformly distributed load, the fixed-end moments are shown in figure below.

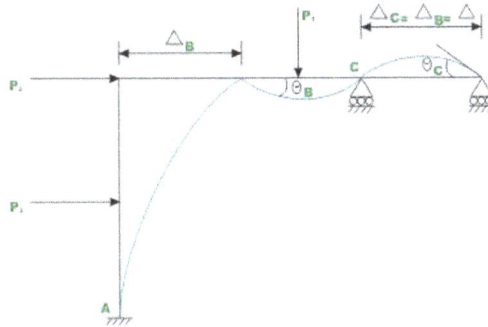

The fixed end moments are required for various load cases. For ease of calculations, fixed end forces for various load cases. In the actual structure end A rotates by $\theta_A$ and end B rotates by $\theta_B$. Now it is required to derive a relation relating $\theta_A$ and $\theta_B$ with the end moments and . Towards this end, now consider a simply supported beam acted by moment $M'_{AB}$ and $M'_{BA}$ at A as shown in figure. The end moment $M'_{AB}$ deflects the beam as shown in the figure. The rotations $\theta'_A$ and $\theta'_B$ are calculated from moment-area theorem.

$$\theta'_A = \frac{M'_{AB}L}{3EI}$$

$$\theta'_B = -\frac{M'_{AB}L}{6EI}$$

Now a similar relation may be derived if only $M'_{BA}$ is acting at end B.

$$\theta''_B = \frac{M'_{BA}L}{3EI} \quad and$$

$$\theta''_A = -\frac{M'_{BA}L}{6EI}$$

Now combining these two relations, we could relate end moments acting at A and B to rotations produced at A and B as,

$$\theta_A = \frac{M'_{AB}L}{3EI} - \frac{M'_{BA}L}{6EI}$$

$$\theta_B = \frac{M'_{BA}L}{3EI} - \frac{M'_{BA}L}{6EI}$$

Solving for $M'_{AB}$ and $M'_{BA}$ in terms of $\theta_A$ and $\theta_B$,

$$M'_{AB} = \frac{2EI}{L}\left(2\theta_A + \theta_B\right)$$

$$M'_{BA} = \frac{2EI}{L}\left(2\theta_B + \theta_A\right)$$

Now writing the equilibrium equation for joint moment at A.

$$M_{AB} = M_{AB}^F + M_{AB}'$$

Similarly writing equilibrium equation for joint B.

$$M_{BA} = M_{BA}^F + M_{BA}'$$

Substituting the value of $M_{AB}'$ from equation $M_{AB}' = \dfrac{2EI}{L}\left(2\theta_A + \theta_B\right)$ in equation $M_{AB} = M_{AB}^F + M_{AB}'$ one obtains,

$$M_{AB} = M_{BA}^F + \frac{2EI}{L}\left(2\theta_A + \theta_B\right)$$

Similarly substituting MBA' from equation $M_{BA}' = \dfrac{2EI}{L}\left(2\theta_B + \theta_A\right)$ in equation $M_{BA} = M_{BA}^F + M_{BA}'$ one obtains,

$$M_{AB} = M_{BA}^F + \frac{2EI}{L}\left(2\theta_B + \theta_A\right)$$

Sometimes one end is referred to as near end and the other end as the far end. In that case, the above equation may be stated as the internal moment at the near end of the span is equal to the fixed end moment at the near end due to external loads plus $\dfrac{2EI}{L}$ times the sum of twice the slope at the near end and the slope at the far end. The above two equations $M_{AB} = M_{BA}^F + \dfrac{2EI}{L}\left(2\theta_A + \theta_B\right)$ and $M_{AB} = M_{BA}^F + \dfrac{2EI}{L}\left(2\theta_B + \theta_A\right)$ simply referred to as slope–deflection equations. The slope-deflection equation is nothing but a load displacement relationship.

## Application of Slope-Deflection Equations to Statically Indeterminate Beams

The procedure is the same whether it is applied to beams or frames. It may be summarized as follows:

1.  Identify all kinematic degrees of freedom for the given problem. This can be done by drawing the deflection shape of the structure. All degrees of freedom are treated as unknowns in slope-deflection method.

2.  Determine the fixed end moments at each end of the span to applied load. The table given at the end of this lesson may be used for this purpose.

3.  Express all internal end moments in terms of fixed end moments and near end, and far end joint rotations by slope-deflection equations.

4.  Write down one equilibrium equation for each unknown joint rotation. For example, at a support in a continuous beam, the sum of all moments corresponding to an unknown joint rotation at that support must be zero. Write down as many equilibrium equations as there are unknown joint rotations.

5. Solve the above set of equilibrium equations for joint rotations.

6. Now substituting these joint rotations in the slope-deflection equations evaluate the end moments.

7. Determine all rotations.

Example:

A continuous beam $ABC$ is carrying uniformly distributed load of 2 kN/m in addition to a concentrated load of 20 kN as shown in Fig.14.5a. Draw bending moment and shear force diagrams. Assume $EI$ to be constant.

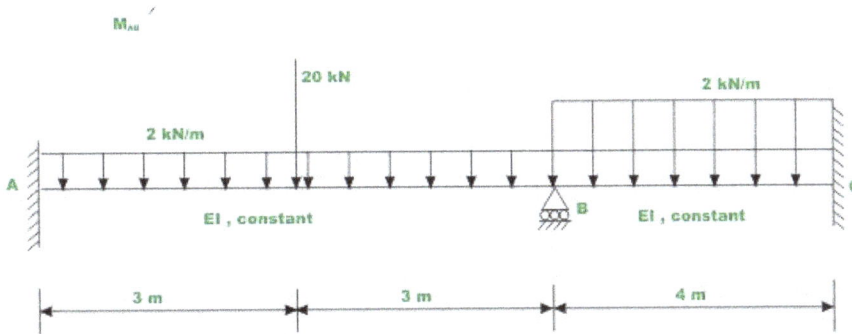

Figure: (a) Example

(a). Degrees of freedom

It is observed that the continuous beam is kinematically indeterminate to first degree as only one joint rotation $\theta_B$ is unknown. The deflected shape /elastic curve of the beam is drawn in Figure. in order to identify degrees of freedom. By fixing the support or restraining the support $B$ against rotation, the fixed-fixed beams area obtained as shown in figure below.

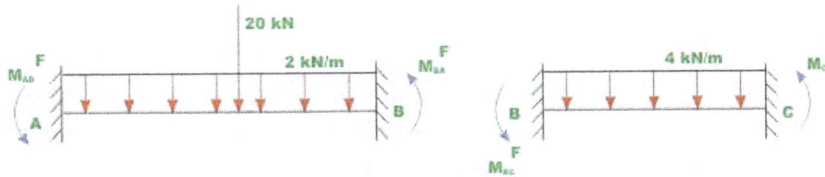

Figure: (b) Restrained structure.

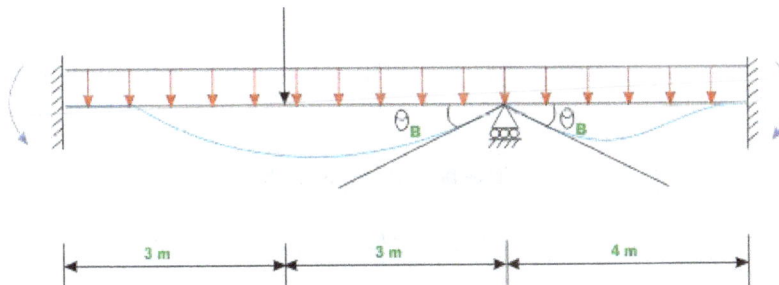

Figure: (c) Elastic curve of the beam with unknown displacement component

(b). Fixed end moments $M_{AB}^F$, $M_{BA}^F$, $M_{BC}^F$ and $M_{CB}^F$ are calculated referring to the figure and following the sign conventions that counter clockwise moments are positive.

$$M_{AB}^F = \frac{2 \times 6^2}{12} + \frac{20 \times 3 \times 3^2}{6^2} = 21 \, \text{kN.m}$$

$$M_{BA}^F = -21 \, \text{kN.m}$$

$$M_{BC}^F = \frac{4 \times 4^2}{12} = 5.33 \, \text{kN.m}$$

$$M_{CB}^F = -5.33 \, \text{KN.m}$$

(c) Slope-deflection equations

Since ends A and C are fixed, the rotation at the fixed supports is zero $\theta_A = \theta_C = 0$. Only one non-zero rotation is to be evaluated for this problem. Now, write slope-deflection equations for span $AB$ and $BC$.

$$M_{AB} = M_{AB}^F + \frac{2EI}{l}(2\theta_A + \theta_B)$$

$$M_{AB} = 21 + \frac{2EI}{6}\theta_B$$

$$M_{AB} = -21 + \frac{2EI}{l}(2\theta_B + \theta_A)$$

$$M_{BA} = -21 + \frac{4EI}{6}\theta_B$$

$$M_{BC} = 5.33 + EI\theta_B$$

$$M_{CB} = -5.33 + 0.5EI\theta_B$$

(d) Equilibrium equations

In the above four equations, the member end moments are expressed in terms of unknown rotation $\theta_B$. Now, the required equation to solve for the rotation $\theta_B$ is the moment equilibrium equation at support B. The free body diagram of support B along with the support moments acting on it is shown in figure. For, moment equilibrium at support B, one must have,

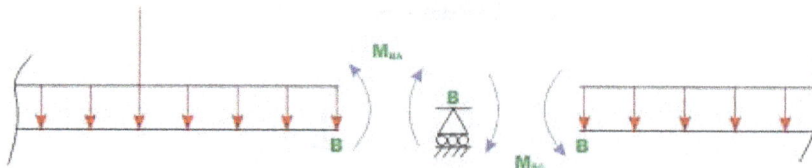

Figure: (d) Free- body diagram of the joint B

$$\sum M_B = 0 \qquad\qquad M_{BA} + M_{BC} = 0$$

Substituting the values of $M_{BA}$ and $M_{BC}$ in the above equilibrium equation,

$$-21 + \frac{4EI}{6}\theta_B + 5.33 + EI\theta_B = 0$$

$$\Rightarrow 1.667\theta_B EI = 15.667$$

$$\theta_B = \frac{9.398}{EI} \cong \frac{9.40}{EI}$$

(e) End moments

After evaluating $\theta_B$, substitute it in equations to evaluate beam end moments. Thus,

$$M_{AB} = 21 + \frac{EI}{3}\theta_B$$

$$M_{AB} = 21 + \frac{EI}{3} \times \frac{9.398}{EI} = 24.133\,\text{kN.m}$$

$$M_{BA} = -21 + \frac{EI}{3}(2\theta_B)$$

$$M_{BA} = -21 + \frac{EI}{3} \times \frac{2 \times 9.4}{EI} = -14.733\,\text{kN.m}$$

$$M_{BC} = 5.333 + \frac{9.4}{EI}EI = 14.733\,\text{kN.m}$$

$$M_{CB} = -5.333 + \frac{9.4}{EI} \times \frac{EI}{2} = -0.63\,\text{kN.m}$$

(f) Reactions

Now, reactions at supports are evaluated using equilibrium equations.

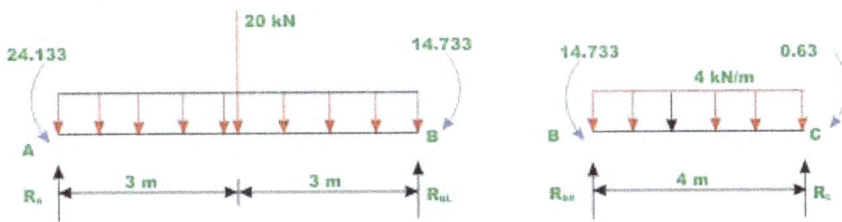

Figure: (e) Free-body diagram of two members

$$R_A \times 6 + 14.733 - 20 \times 3 - 2 \times 6 \times 3 - 24.133 = 0$$

$$R_A = 17.567\,\text{kN}(\uparrow)$$

$$R_{BL} = 16 - 1.567 = 14.433\,\text{kN}(\uparrow)$$

$$R_{BR} = 8 + \frac{14.733 - 0.63}{4} = 11.526\,\text{kN}(\uparrow)$$

$$_s R_C = 8 + 3.526 = 4.47\,\text{kN}(\uparrow)$$

The shear force and bending moment diagrams are shown in figure.

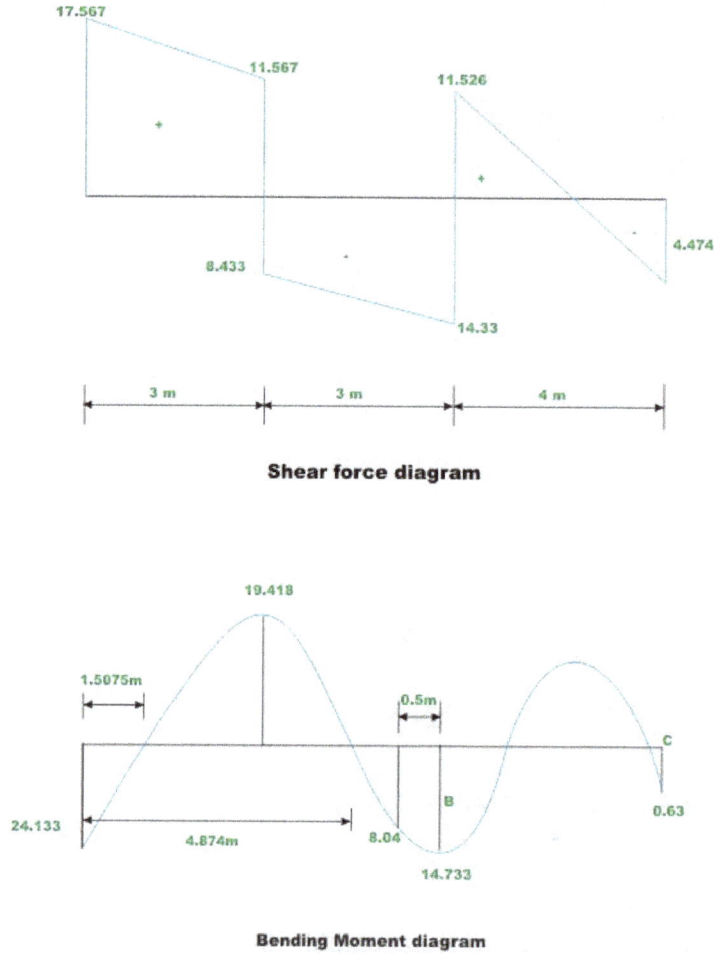

**Shear force diagram**

**Bending Moment diagram**

Figure (f) Sheare force and bending moment diagram of continuous beam ABC

Example:

Draw shear force and bending moment diagram for the continuous beam loaded as shown in figure.The relative stiffness of each span of the beam is also shown in the figure.

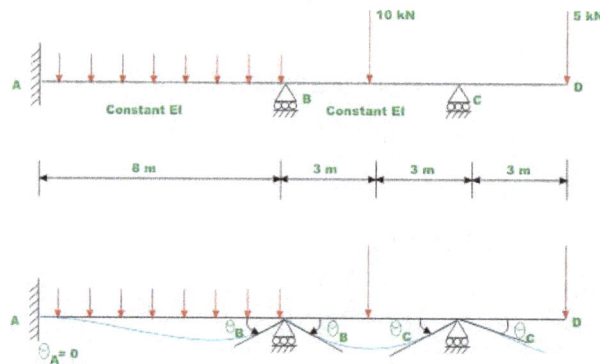

Figure: Continuous beam

For the cantilever beam portion $CD$, no slope-deflection equation need to be written as there is no internal moment at end $D$. First, fixing the supports at $B$ and $C$, calculate the fixed end moments for span $AB$ and $BC$. Thus,

$$M_{AB}^F = \frac{3 \times 8^2}{12} = 16\,\text{kN.m}$$

$$M_{BA}^F = -16\,\text{kN.m}$$

$$M_{BC}^F = \frac{10 \times 3 \times 3^2}{6^2} = 7.5\,\text{kN.m}$$

$$M_{CB}^F = -7.5\,\text{kN.m}$$

In the next step write slope-deflection equation. There are two equations for each span of the continuous beam.

$$M_{AB} = 16 + \frac{2EI}{8}(\theta_B) = 16 + 0.25\theta_B EI$$

$$M_{BA} = -16 + 0.5_B\,EI$$

$$M_{BC} = 7.5 + \frac{2 \times 2EI}{6}(2\theta_B + \theta_C) = 7.5 + 1.334EI\theta_B + 0.667EI\theta_C$$

$$M_{CB} = -7.5 + 1.334EI\theta_C + 0.667EI\theta_B$$

Equilibrium equations

The free body diagram of members $AB$, $BC$ and joints $B$ and $C$ are shown in figure. One could write one equilibrium equation for each joint $B$ and $C$.

Figure: Free body diagrams of joints B and C along with members

Support B,

$$\sum M_B = 0 \qquad\qquad M_{BA} + M_{BC} = 0$$

$$\sum M_C = 0 \qquad\qquad M_{CB} + M_{CD} = 0$$

We know that $M_{CD} = 15\,\text{kN.m}$.

$$\Rightarrow M_{CB} = -15\,\text{kN.m}$$

Substituting the values of $M_{CB}$ and $M_{CD}$ in the above equations for $M_{AB}, M_{BA}, M_{CB}$ and $M_{CB}$ we get,

$$\theta_B = \frac{24.5}{3.001} = 8.164$$

$$\theta_C = 9.704$$

Substituting $\theta_B$, $\theta_C$ in the slope-deflection equations, we get,

$$M_{AB} = 16 + 0.25EI\theta_B = 16 + 0.25EI \times \frac{8.164}{EI} = 18.04\,\text{kN.m}$$

$$M_{BA} = -16 + 0.5EI\theta_B = -16 + 0.5EI \times \frac{8.164}{EI} = -11.918\,\text{kN.m}$$

$$M_{BC} = 7.5 + 1.334EI \times \frac{8.164}{EI} + 0.667EI\left(\frac{9.704}{EI}\right) = 11.918\,\text{kN.m}$$

$$M_{CB} = -7.5 + 0.667EI \times \frac{8.164}{EI} + 1.334EI\left(-\frac{9.704}{EI}\right) = -15\,\text{kN.m}$$

Reactions are obtained from equilibrium equations,

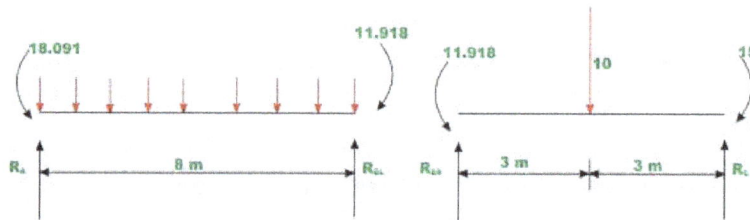

Figure: Computation of reactions

$$R_A \times 8 - 18.041 - 3 \times 8 \times 4 + 11.918 = 0$$
$$R_A = 12.765\,\text{kN}$$
$$R_{BR} = 5 - 0.514\,\text{kN} = 4.486\,\text{kN}$$
$$R_{BL} = 11.235\,\text{kN}$$
$$R_C = 5 + 0.514\,\text{KN} = 5.514\,kN$$

The shear force and bending moment diagrams are shown in figure.

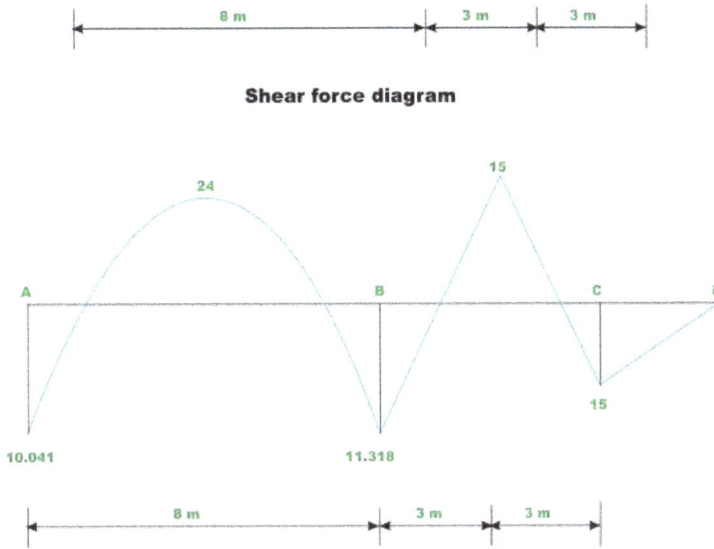

Figrue: Shear force and bending moment diagram

For ease of calculations, fixed end forces for various load cases.

## Slope Deflection Equations

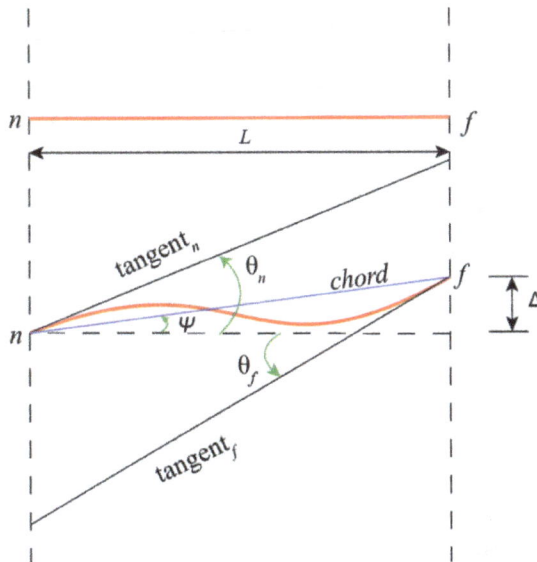

The equations found for use in the Slope Deflection Method connect the rotations and displacements of a beam to the moments that are induced at its ends through forces. The elastic curve of a beam under a set of arbitrary forces is used to compare the differences in rotation of the beam to the angle of the chord.

$$M_{nf} = \frac{2EI}{L}(2\theta_n + \theta_f - 3\psi) + M'_n fM \sim$$

Where $M_{nf}$ is the Moment acting at near end denoted as subscript n of member while subscript f represent the far/other end of the member, $M'$ is Fixed-End Moments. $\theta_n$ and $\theta_f$ are the near end and far end rotations of the member with respect to horizontal position of the member. $\psi$ represent the rotation of member's chord under consideration. It is given as $\psi = \dfrac{\Delta}{L}$ where, $\Delta$ is deformation assumed to be small. All the parameters are shown in figure 1. Eq. 1 is valid only for prismatic members composed of linearly elastic member and subjected to small deformations also, axial and shear forces are neglected.

## Analysis of Continuous Beams

The method for analysis of continuous beams is adapted from the textbook, Structural Analysys by Kassimali.

### Step 1

Evaluate the Degree of Freedom. In the case of a continuous beam, the Degree of Freedom will be the number of unknown reactions.

### Step 2

Formulate Equilibrium Equations. From Step 1, the number of Equilibrium Equations will be known as it is equal to the Degree of Freedom of the system.

### Step 3

Calculate fixed-end moments for each member in the system. In most cases, the fixed-end moments will be one of the unknown reactions of the beam.

### Step 4

Calculate the chord rotations for the member under consideration adjacent to support with support settlement using relation $\psi = \dfrac{\Delta}{L}$. Care must be taken for sign convention clockwise rotation is negative and vice versa.

### Step 5

Using $M_{nf} = \dfrac{2EI}{L}(2\theta_n + \theta_f - 3\psi) + M'_n fM \sim$ formulate the slope deflection equation, for each member there will be two equations.

### Step 6

Substitute slope deflection equation from step 5 into the equilibrium equation derived in step 2 and evaluate unknown θ joint rotations.

### Step 7

Substitute θ value in slope deflection equations from step 5 and calculate members end moments.

Note * Counter clockwise moments are consider positive.

Step 8

To cross check solution from step 7, the member end moment muse be substituted into the Equilibrium Equation from step 2.

Step 9

Calculate support reactions.

Step 10

Draw a free body diagram for the beam in question.

Step 11

Draw Shear Force and Bending Moment diagrams for the beam.

## Unit Dummy Force Method

The Unit dummy force method provides a convenient means for computing displacements in structural systems. It is applicable for both linear and non-linear material behaviours as well as for systems subject to environmental effects, and hence more general than Castigliano's second theorem.

### Discrete Systems

Consider a discrete system such as trusses, beams or frames having members interconnected at the nodes. Let the consistent set of members' deformations be given by $r_{N \times 1}$, which can be computed using the member flexibility relation. These member deformations give rise to the nodal $r_{N \times 1}$, which we want to determine.

We start by applying N virtual nodal forces $R^*_{N \times 1}$, one for each wanted $r$, and find the virtual member forces $Q^*_{M \times 1}$ that are in equilibrium $R^*_{N \times 1}$,:

$$Q^*_{M \times 1} = B_{M \times N} R^*_{N \times 1}$$

In the case of a statically indeterminate system, matrix $B$ is not unique because the set of $Q^*_{M \times 1}$ that satisfies nodal equilibrium is infinite. It can be computed as the inverse of the nodal equilibrium matrix of any primary system derived from the original system.

Imagine that internal and external virtual forces undergo, respectively, the real deformations and displacements; the virtual work done can be expressed as:

- External virtual work: $R^{*T} r$
- Internal virtual work: $Q^{*T} q$

According to the virtual work principle, the two work expressions are equal:

$$R^{*T}r = Q^{*T}q$$

Substitution of $(Q^*_{M \times 1} = B_{M \times N} R^*_{N \times 1})$ gives,

$$R^{*T}r = R^{*T}B^T q$$

Since $R^*$ contains arbitrary virtual forces, the above equation gives,

$$r = B^T q$$

It is remarkable that the computation in $r = B^T q$ does not involve any integration regardless of the complexity of the systems, and that the result is unique irrespective of the choice of primary system for $B$. It is thus far more convenient and general than the classical form of the dummy unit load method, which varies with the type of system as well as with the imposed external effects. On the other hand, it is important to note that $r = B^T q$ is for computing displacements or rotations of the nodes only. This is not a restriction because we can make any point into a node when desired.

Finally, the name unit load arises from the interepretation that the coefficients $B_{i,j}$ in matrix $B$ are the member forces in equilibrium with the unit nodal force $R^*_j = 1$, , by virtue of $Q^*_{M \times 1} = B_{M \times N} R^*_{N \times 1}$.

## General Systems

For a general system, the unit dummy force method also comes directly from the virtual work principle. Figure shows a system with known actual deformations $\epsilon$. These deformations, supposedly consistent, give rise to displacements throughout the system. For example, a point $A$ has moved to $A'$, and we want to compute the displacement $r$ of $A$ in the direction shown. For this particular purpose, we choose the virtual force system in figure which shows:

- The unit force R* is at A and in the direction of r so that the external virtual work done by R* is, noting that the work done by the virtual reactions in (b) is zero because their displacements in (a) are zero:

  $$R^* \times r = 1 \times r : \text{The desired displacement}$$

- The internal virtual work done by the virtual stresses is

- $$\int_V \epsilon^T \acute{o}^* dV$$

where the virtual stresses $\acute{o}^*$ must satisfy equilibrium everywhere.

Equating the two work expressions gives the desired displacement:

$$1 \times r = \int_V \epsilon^T \sigma^* dV$$

# Flexibility Method

The flexibility method is categorized as a force method because it regards the forces in the structure as the primary unknowns. In contrast to displacement methods, in which equilibrium equations are postulated to determine the unknown deformations, the core of the flexibility method is the formulation of compatibility equations. The key steps of the method are:

1. Determine the structure's degree of static indeterminacy.

2. Make the structure statically determinate by introducing releases.

3. Establish compatibility equations to ensure continuity at the releases.

4. Solve the compatibility equations and draw the final section force diagram (M, V, N) for each member.

Usually, the number of releases equals the degree of static indeterminacy. Exceptions are when some forces are known in particular load cases. For trusses, it is common to remove supports and cut members to introduce releases. For frame members, it is common to remove supports, release bending moments by introducing hinges, and sometimes to release shear and axial forces.

The displacements or rotations that take place along the releases are called "gaps." The gap that opens due to the actual loads on the structure is called $\Delta_{io}$, where the index $i$ indicates the location of the release, while o says it is due to the actual loads. Furthermore, let $\Delta_{ij}$ denote the gap that opens at location i due to a unit load at location $j$. These are called flexibility coefficients because they reveal the force it takes to increase/decrease the gap. Compatibility equations serve the purpose of closing the gaps. In index notation, they read:

$$\Delta_{io} + \Delta_{ij} \cdot x_j = 0$$

where $x_j$ are the unknown forces (axial force, bending moment, or shear force) at the releases. In matrix notation, the compatibility equations read:

$$d + fx = 0$$

where $d$ is the vector of gaps due to the actual forces, $f$ is the matrix of flexibility coefficients, and x is the vector of unknown forces at the releases. For a structure with one degree of static indeterminacy, where a force at location "A" is selected as the redundant, the compatibility equation reads:

$$\Delta_{AO} + \Delta_{AA} \cdot x_A = 0$$

For a structure with two degrees of static indeterminacy, where a force at location "A" and a force at location "B" are selected as redundants, the compatibility equations read:

$$\Delta_{AO} + \Delta_{AA} \cdot x_A + \Delta_{AB} \cdot x_B = 0$$
$$\Delta_{BO} + \Delta_{BA} \cdot x_A + \Delta_{BB} \cdot x_B = 0$$

In summary, there are two key steps to establish the compatibility equations: First select releases that make the structure statically determinate and then determine the deformations $\Delta_{io}$ and $\Delta_{ij}$. The first step is addressed on an ad hoc basis. This is the reason why the flexibility method is rarely implemented on the computer; it is difficult to set up a generic algorithm to select releases for any statically indeterminate structure. The second step is usually solved by the unit virtual load method. I.e., the deformations $\Delta_{io}$ and $\Delta_{ij}$ are determined by applying unit virtual forces along the releases. In this document, the section force diagrams due to a unit force along release number $j$ are denoted $M_j$, $V_j$, and $N_j$. Upon solving the compatibility equations in Equation $\Delta_{io} + \Delta_{ij} \cdot x_j = 0$ for $x_j$ the final section force diagrams are determined by combining the diagrams for the statically determinate auxiliary structure:

$$M = M_O + M_A \cdot x_A + M_B \cdot x_B$$
$$V = V_O + V_A \cdot x_A + V_B \cdot x_B$$
$$N = N_O + N_A \cdot x_A + N_B \cdot x_B$$

where $M_0$ is the bending moment diagram for the statically determinate structure subjected to the applied loads, $\delta M_A$ is the bending moment diagram for the statically determinate structure due to a unit virtual load at $A$, and so forth.

## Settlements and Changes in Member Lengths

With one exception, the effect of settlements and changes in member lengths are included in the left-most terms in the left-hand side of the compatibility equations. Specifically, the additional "gap openings" due to settlements and length changes are added to $\Delta_{io}$. These are conveniently computed by the unit virtual load method, as described towards the end of the document on that method. The one exception appears when the settlement takes place exactly at the redundant. For example, if a support reaction is selected as the redundant and that support settles, then this settlement is placed at the right-hand side of the compatibility equation instead of zero. If the settlement is in the same direction as the positive direction of the redundant then the settlement value is entered with a positive sign in the right-hand side.

An interesting case of member length change is post-tensioning. As an example, think of a truss member that is subjected to post-tensioning by some bar tensioning mechanism, after the truss is built. Suppose the forces and deformations in the structure are sought under these circumstances. The flexibility method addresses this problem as follows:

1.  Consider the structure without any external loads;

2.  Make the structure statically determinate by introducing cuts, not necessarily in the post-tensioned members;

3.  Compute the gaps $\Delta_{io}$ due to a unit shortening of the post-tensioned member by the unit virtual load method;

4.  As usual, determine the flexibility coefficients, establish the compatibility equations, solve them, and draw the final section force diagram ($M$, $V$, $N$) for each member; this provides the relationship between the unit member shortening and the member forces;

5.  If the deformations are sought, re-analyse the structure for a unit load at the location where the displacement or rotation is sought and apply the unit virtual load method; this provides the relationship between the unit member shortening and the sought deformation.

# Shear and Moment Diagram

Shear force diagram and bending moment diagrams are illustrations to describe the alterations in shear force and bending moments over the length of the beam. These diagrams can be utilized to analyse the failure of the structure with the given inputs like loads, structure material, and shape. Hence, SFD and BMD reduce the probability of the structure's failure. Also with the implementation of conjugate beam method or moment area method, the deflection of the beam can be resolved.

## Shear Force Diagram

To draw a shear force diagram, the below series of steps are adopted.

1. Determine the internal forces (reactions) at the supports and use sign conventions. Take the upward force as positive and downward force as negative.

2. Starting from left, plot the shear force values with respect to the point of action on the beam.

3. The curvature for joining these points depends upon the type of load as shown in figure below.

| Load | Slope for shear force | Slope for bending Moment |
|---|---|---|
| P | Constant | Linear |
| Uniformly distributed load | Linear | Parabolic |
| Uniformly varying load | Parabolic | Cubic |

Figure: Slopes for various types of loads

The shear force diagram should start and end at x-axis (SF equals to zero) at the ends. Maximum bending moment exists at the location where the SFD crosses the axis.

## Bending Moment Diagram

Bending moment diagram is drawn using the following steps.

1. Initially convert the uniformly distributed loads (UDL) anduniformlyvarying loads (UVL) acting on the beam into point loads.

2. Starting from left, calculate the value of bending moment for the loads acting on the beam.

3. Plot the values of bending moment on the corresponding location on the beam.

4. Join these points using curvatures depending upon the type of load as shown in figure above.

An example of SFD and BMD for a simply supported beam is shown in figure below.

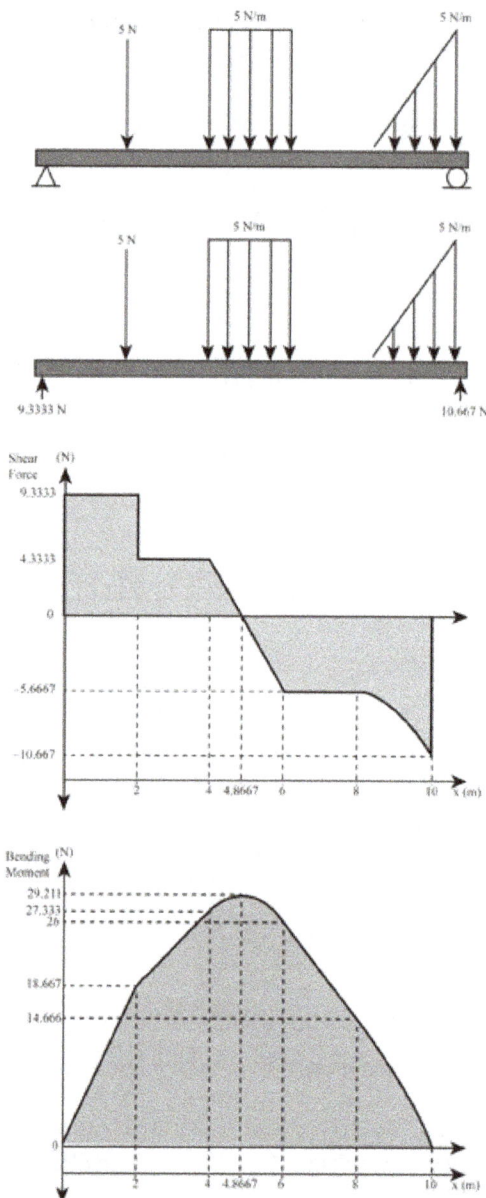

# Moment-area Theorem

The moment-area method is one of the most effective methods for obtaining the bending displacement in beams and frames. In this method, the area of the bending moment diagrams is utilized for computing the slope and or deflections at particular points along the axis of the beam or frame. Two theorems known as the moment area theorems are utilized for calculation of the deflection. One theorem is used to calculate the change in the slope between two points on the elastic curve. The other theorem is used to compute the vertical distance (called tangential deviation) between a point on the elastic curve and a line tangent to the elastic curve at a second point.

Consider figure below showing the elastic curve of a loaded simple beam. On the elastic curve tangents are drawn on points $A$ and $B$. Total angle between the two tangents is denoted as $\Delta\theta_{AB}$. In order to find out $\Delta\theta_{AB}$, consider the incremental change in angle $d\theta$ over an infinitesimal segment $dx$ located at a distance of from point $B$. The radius of curvature and bending moment for any section of the beam is given by the usual bending equation.

$$\frac{M}{I} = \frac{E}{R}$$

where R is the radius of curvature; E is the modulus of elasticity; I is the moment of inertia; and M denotes the bending moment.

The elementary length $dx$ and the change in angle $d\theta$ are related as,

$$dx = d\theta \times R$$

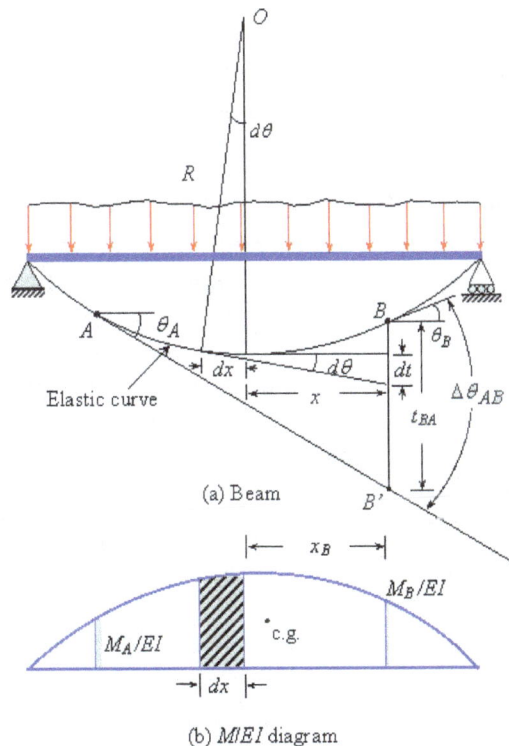

(a) Beam

(b) $M/EI$ diagram

Substituting R from Eq. $dx = d\theta \times R$ in Eq. $\dfrac{M}{I} = \dfrac{E}{R}$

$$d\theta = \frac{M}{EI}dx$$

The total angle change $\Delta\theta_{AB}$ can be obtained by integrating Eq. $d\theta = \dfrac{M}{EI}dx$ between points A and B which is expressed as,

$$\Delta\theta_{AB} = \int_A^B dx = \int_A^B \frac{M}{EI}dx$$

or,

$$\theta_B - \theta_A = \text{Area of M/EI diagram between A and B}$$

The difference of slope between any two points on a continuous elastic curve of a beam is equal to the area under the M/EI curve between these points.

The distance dt along the vertical line through point B is nearly equal to

$$dt = x \times d\theta$$

Integration of dt between points A and B yield the vertical distance $t_{BA}$ between the point B and the tangent from point A on the elastic curve. Thus,

$$t_{BA} = \int_A^B xd\theta = \int_A^B \frac{Mx}{EI}dx$$

since the quantity $Md_x/EI$ represents an infinitesimal area under the M /EI diagram and distance x from that area to point B, the integral on right hand side of Eq. can be interpreted as moment of the area under the M/EI diagram between points A and B about point B . This is the second moment area theorem.

If A and B are two points on the deflected shape of a beam, the vertical distance of point B from the tangent drawn to the elastic curve at point A is equal to the moment of bending moment diagram area between the points A and B about the vertical line from point B , divided by EI .

Sign convention used here can be remembered keeping the simply supported beam of figure in mind. A sagging moment is the positive bending moment diagram and has positive area. Slopes are positive if measured in the anti-clockwise direction. Positive deviation $t_{BA}$ indicates that the point B lies above the tangent from the point A .

Example: Determine the end slope and deflection of the mid-point C in the beam shown below using moment area method.

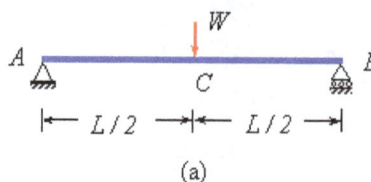

(a)

(b)

(c)

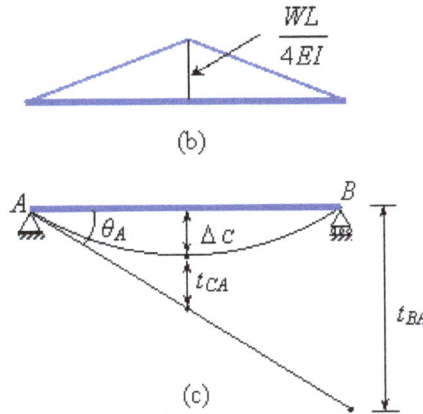

Solution: The M/EI diagram of the beam is shown in figure (a). The slope at A, $\theta_A$ can be obtained by computing the $t_{BA}$ using the second moment area theorem i.e.

$$\theta_A \times L = t_{BA}$$

$$\theta_A = \frac{1}{L} \times \left( \frac{1}{2} \times \frac{WL}{4EI} \times L \times \frac{L}{2} \right) = \frac{WL^2}{16EI} \; (clockwise\,direction)$$

The slope at B can be obtained by using the first moment area theorem between points A and B i.e.

$$\theta_B - \theta_A = \Delta\theta_{AB}$$

$$\theta_B - \theta_A = \frac{1}{2} \times \frac{WL}{4EI} \times L = \frac{WL^2}{8EI}$$

$$\theta_B = \frac{WL^2}{8EI} - \frac{WL^2}{16EI} = \frac{WL^2}{16EI} \; (anti-clockwise)$$

(It is to be noted that the $\theta_A = \frac{WL^2}{16EI}$. The negative sign is because of the slope being in the clockwise direction. As per sign convention a positive slope is in the anti-clockwise direction).

The deflection at the centre of the beam can be obtained with the help of the second moment area theorem between points A and C i.e.

$$\theta_A \times \frac{L}{2} = \Delta_C + t_{CA}$$

$$\frac{WL^2}{16EI} \times \frac{L}{2} = \Delta_C + \left( \frac{1}{2} \times \frac{WL}{4EI} \times \frac{L}{2} \times \frac{L}{6} \right)$$

$$\Delta_C = \frac{WL^3}{48EI} \; (download\;direction)$$

Example: Using the moment area method, determine the slope at B and C and deflection at C of the cantilever beam as shown in figure: The beam is subjected to uniformly distributed load over entire length and point load at the free end.

Solution: The moment curves produced by the concentrated load, $W$ and the uniformly distributed load, $w$ are plotted separately and divided by $EI$. This results in the simple geometric shapes in which the area and locations of their centroids are known.

Since the end A is fixed, therefore, $\theta_A = 0$ Applying the first moment-area theorem between points A and C.

$$\theta_C - \theta_A = \Delta\theta_{AC}$$

$$\theta_C - \theta_A = -\left(\frac{1}{2} \times L \times \frac{WL}{EI} \times \frac{1}{3} \times L \times \frac{wL^2}{2EI}\right) \text{ (negative sign is due to hogging moment)}$$

$$\theta_C = -\left(\frac{WL^2}{2EI} + \frac{wL^3}{6EI}\right) \text{ (clockwise direction)}$$

The slope at B can be obtained by applying the first moment area theorem between points B and C i.e.

$$\theta_C - \theta_B = \Delta\theta_{BC}$$
$$\theta_B = \theta_C - \Delta\theta_{BC}$$
$$\theta_B = -\left(-\frac{wL^2}{2EI} + \frac{wL^3}{2EI}\right) - \left(\frac{1}{2} \times \frac{L}{2} \times \frac{WL}{2EI} - \frac{1}{3} \times \frac{L}{2} \times \frac{wL^2}{8EI}\right)$$

$$\theta_B = -\left(\frac{3wL^2}{8EI} + \frac{7wL^3}{48EI}\right) \text{ (clockwise dirction)}$$

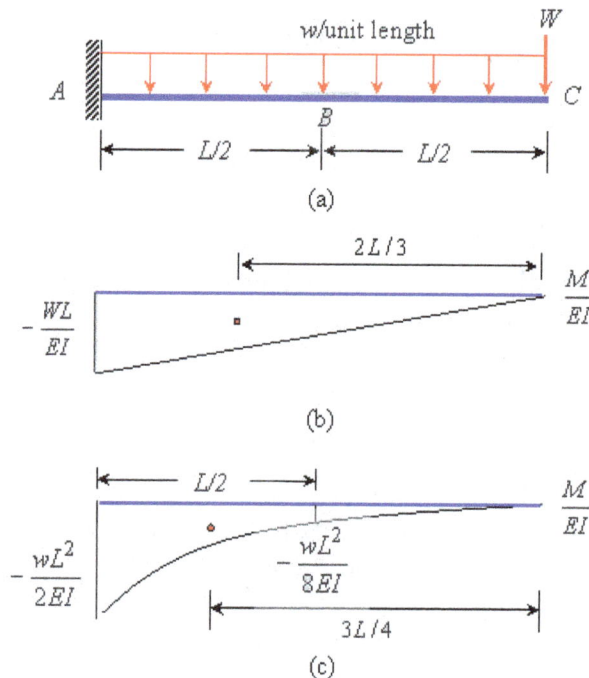

(a)

(b)

(c)

tangent at $A$

(d)

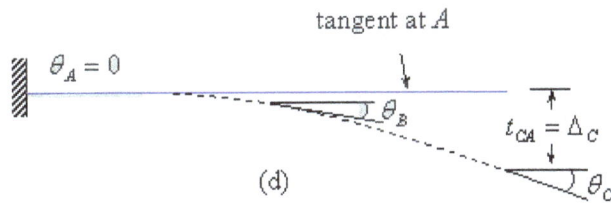

The deflection at C is equal to the tangential deviation of point C from the tangent to the elastic curve at A.

$\Delta_C = t_{CA}$ = moment of areas under M/EI curves between A and C in figures above (b) and (c) about C

$$= \frac{1}{2} \times L \times \frac{WL}{EI} \times \frac{2L}{3} + \frac{1}{3} \times L \times \frac{wL^2}{2\,EI} \times \frac{3L}{4}$$

$$= \frac{WL^3}{3\,EI} + \frac{wL^4}{8\,EI} \quad (download\ direction)$$

Example: Determine the end-slopes and deflection at the centre of a non-prismatic simply supported beam. The beam is subjected to a concentrated load at the centre.

Solution: The M/EI diagram of the beam is shown in figure.

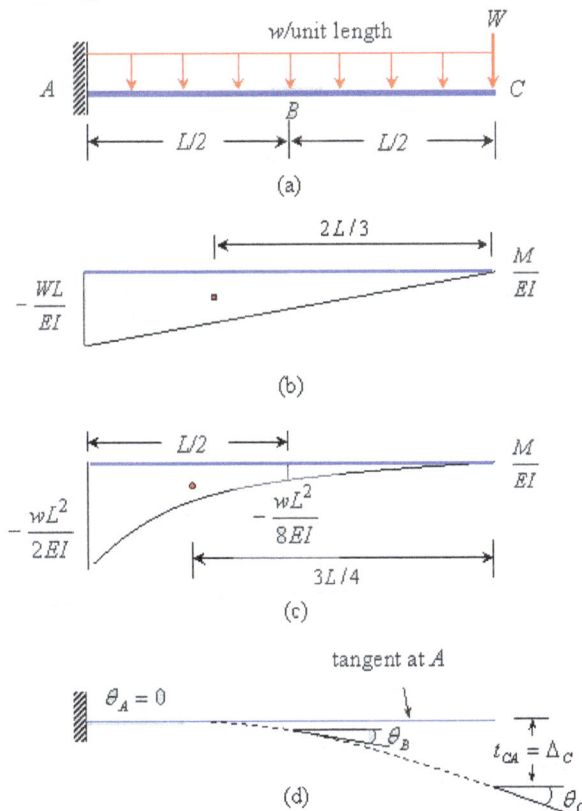

(a)

(b)

(c)

tangent at $A$

(d)

Applying second moment-area theorem between points A and B,

$$t_{BA} = \frac{1}{2} \times \frac{L}{2} \times \frac{PL}{4EI}\left(\frac{L}{2} + \frac{L}{6}\right) + \frac{1}{2} \times \frac{PL}{8EI} \times \frac{L}{2} \times \left(\frac{2}{3} \times \frac{L}{2}\right)$$

$$-\theta_A L = \frac{PL^3}{24EI} + \frac{PL^3}{96EI} = \frac{5PL^3}{96EI}$$

$$\theta_A = -\frac{5PL^2}{96EI} \; (clockwise\,direction)$$

Applying first moment area theorem between A and C.

$$\theta_C - \theta_A = \frac{1}{2} \times \frac{L}{2} \times \frac{PL}{4EI}$$

$$\theta_C = \frac{PL^2}{16EI} - \frac{5PL^2}{96EI} = \frac{PL^2}{96EI} \; (anti-clockwise\,direction)$$

Applying second moment area theorem between A and C.

$$t_{CA} = \frac{PL^2}{16EI} \times \frac{L}{6} = \frac{PL^3}{96EI}$$

$$\left(\theta_A \times \frac{L}{2}\right) - \Delta_C = \frac{PL^3}{96EI}$$

$$\Delta_C = \frac{5PL^2}{96EI} \times \frac{L}{2} - \frac{PL^3}{96EI} = \frac{PL^3}{64EI} \; (download\;direction)$$

Example: Determine the slope and deflection at the hinge of the beam shown in the figure.

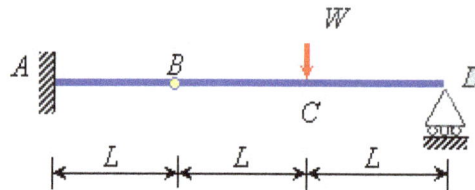

Solution: The bending moment diagram is shown in figure(b).

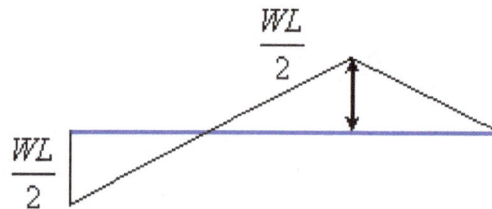

Since the end A is fixed, therefore, $\theta_A = 0$. Applying the first moment-area theorem between points A and B.

$$\theta_{BA} - \theta_A = -\frac{1}{2} \times \frac{WL}{2} \times L \times \frac{1}{EI}$$

$$\theta_{BA} = -\frac{WL^2}{4EI} \ (clockwise \ direction)$$

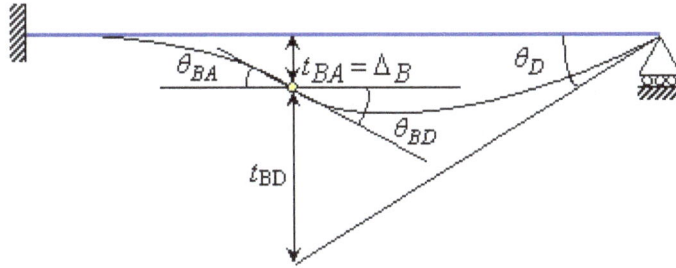

Applying second moment area theorem between points A and B,

$$\Delta_B = t_{BA} = \frac{1}{2} \times \frac{WL}{2} \times L \times \frac{1}{EI} \times \frac{2L}{3}$$

$$= \frac{WL^3}{6EI} \ (downward \ direction)$$

Applying second moment area theorem between points B and D,

$$\theta_D \times 2L = (t_{BA} + t_{BD})$$

$$\theta_D = \frac{1}{2L} \times \left( \frac{WL^3}{6EL} + \frac{1}{2} \times \frac{WL}{2} \times 2L \times \frac{WL}{EL} \times L \right)$$

$$\theta_D = \frac{WL^2}{3EI} (anti\text{-}clockwise \ direction)$$

From the first moment area theorem between points B and D,

$$\theta_D - \theta_{BD} = \frac{1}{2} \times \frac{WL}{2} \times 2L \times \frac{1}{EL}$$

$$\theta_{BD} = \frac{WL^2}{3EI} - \frac{WL^2}{2EI}$$

$$\theta_{BD} = -\frac{WL^2}{6EI} (clockwise \ direction)$$

Example: Determine the vertical deflection and slope of point C of the rigid-jointed plane frame shown in the figure.

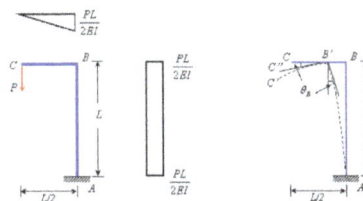

Solution: The M/EI and deflected shape of the frame are shown in the figures (a) and (b), respectively. As the point A is fixed implying that $\theta_A = 0$. Applying first moment area theorem between points A and B,

$$\theta_A - \theta_B = -\frac{PL}{2EI} \times L \left( \text{looking from the left side} \right)$$

$$\theta_B = \frac{PL^2}{2EI} \quad \text{(anti-clockwise direction)}$$

Applying second moment area theorem between points B and C,

$$t_{CB} = -\frac{1}{2} \times \frac{PL}{2EI} \times \frac{L}{2} \times \left( \frac{2}{3} \times \frac{L}{2} \right) = -\frac{PL^3}{24EI}$$

The vertical displacement of point C,

$$\Delta_C = -\theta_B \times \frac{L}{2} + t_{CB}$$

$$\Delta_C = -\theta_B \times \frac{L}{2} + t_{CB} \left( \text{downward direction} \right)$$

Applying first moment area theorem between point B and C,

$$\theta_B - \theta_C = -\frac{1}{2} \times \frac{PL}{2EI} \times L/2$$

$$\theta_C = \frac{5PL^2}{8EI} \left( \text{anti-clockwise direction} \right)$$

## References

- Basics-of-structural-analysis-44070: brighthubengineering.com, Retrieved 21 May 2018

- Structural-theory, building-design-and-construction-handbook: civilengineeringx.com, Retrieved 29 April 2018

- Engineers-and-structural-dynamics: istructe.org, Retrieved 12 July 2018

- Moving-loads, mechanics-and-strength-of-materials: mathalino.com, Retrieved 06 March 2018

- Shear-and-moment-diagrams-5: chegg.com, Retrieved 25 June 2018

# Permissions

# Index